18

TESI

THESES

tesi di perfezionamento in Fisica sostenuta il 22 gennaio 2013

Giovanni Petrucciani
CERN
Geneva
Switzerland

The Search for the Higgs Boson at CMS

Giovanni Petrucciani

The Search
for the Higgs Boson
at CMS

EDIZIONI
DELLA
NORMALE

ISBN 978-88-7642-481-6
e-ISBN 978-88-7642-482-3

Contents

Introduction

The quest for the standard model Higgs boson

Since its first formulations in the 1960s, the standard model [1, 2, 3] has been extremely successful in describing the physics of all known elementary particles, and has been tested to remarkable accuracy in several generations of experiments. The foundation of the standard model is the existence of a hidden symmetry of the electromagnetic and weak interactions, which is spontaneously broken by the presence in the vacuum of a non-vanishing Higgs boson field [4, 5, 6, 7, 8, 9]. As a consequence of the electroweak symmetry breaking, the weak gauge bosons W and Z acquire masses through their interaction with the Higgs boson field, and their dynamic changes; the masses of quarks and leptons also arise from the same mechanism.

After the symmetry breaking, the Higgs boson field manifests itself as a massive neutral scalar particle H, the dynamics of which are completely predicted by the theory as function of the unknown Higgs boson mass m_H, and of the known masses and couplings to fermions and gauge bosons [10, 11]. The experimental searches for this particle at the Large Electron Positron collider have yielded negative results, thereby excluding the whole kinematic range $m_H < 114.4 \, \mathrm{GeV}/c^2$ at 95% confidence level [12]. Likewise, no evidence of the Higgs boson has been found at the Tevatron collider, and recent analyses of the data collected by the CDF and D0 experiments have ruled out the narrow m_H region 147–$179 \, \mathrm{GeV}/c^2$ at 95% confidence level [13, 14]. In addition to the direct searches, phenomenological considerations bound the Higgs boson mass to be below about 1 TeV, and indirect constraints from precision electroweak measurements favour the hypothesis of a light Higgs boson, $m_H < 158 \, \mathrm{GeV}/c^2$ at 95% confidence level [15].

One of the main goals of the physics programme at the Large Hadron Collider (LHC) [16] is to provide a definitive answer about the existence of the standard model Higgs boson, and more in general about the mechanism of the electroweak symmetry breaking, by significantly expanding both the energy reach and the collision rate with respect to the Tevatron collider. The two general purpose experiments, CMS [17] and ATLAS [18], have been designed with the detection of a Higgs boson among their primary targets.

Searches for a Higgs boson at the LHC
The relevant experimental signatures in the searches for the standard model Higgs boson are determined by the production cross sections and decay branching fractions, dependent on the Higgs boson mass hypothesis, and by the level of standard model backgrounds and the observable Higgs boson mass resolution. At LHC energies, the four main Higgs boson production mechanisms are: (i) gluon fusion through a quark loop, $gg \rightarrow H$, with the largest cross section; (ii) vector boson fusion $qq \rightarrow V^*V^*qq \rightarrow Hqq$, with a cross section smaller by about an order of magnitude but a characteristic di-jet topology; (iii) associated production with a W or Z boson, $q\bar{q} \rightarrow VH$, with an even smaller production rate, but experimentally more accessible when selecting leptonic or invisible W and Z decay modes; (iv) associated production with a $t\bar{t}$ pair, with the smallest rate and a challenging multi-jet final state.

The dominant decay channel for a light Higgs boson is into $b\bar{b}$ pairs, but the $O(10^7)$ larger cross section for the background QCD production of $b\bar{b}$ pairs makes the detection of a signal extremely challenging in practice; present searches are confined to the somewhat more accessible $q\bar{q} \rightarrow VH$ production mode. Conversely, the decay channel into photon pairs has a branching fraction of only $O(0.1\%)$, but the much cleaner experimental signature make this channel the most sensitive one at low masses. An intermediate ground is offered by the $H \rightarrow \tau\tau$ decay, with a $O(5\%)$ branching fraction.

The decays of the Higgs boson into four fermions, mediated by pairs of real or virtual W or Z bosons, become increasingly important for Higgs boson masses above about $125\,\text{GeV}/c^2$, and dominant above the threshold of $2m_W$. The fully leptonic final states $H \rightarrow WW \rightarrow 2\ell\,2\nu$ and $H \rightarrow ZZ \rightarrow 4\ell$, where ℓ stands for an electron or muon, are the two golden modes for searches at intermediate Higgs boson masses. The former is characterized by a large branching fraction, but the presence of two neutrinos in the final state does not allow the kinematics of the event to be fully reconstructed. The main challenges in this search arise then from the need to control all the background processes involved: $W + \text{jets}$ with jets misidentified as leptons, $Z + \text{jets}$ with mis-measured jets causing an imbalance in the transverse energy flow similar the one from undetected neutrinos, $t\bar{t}$ with one of the two b-jets from $t \rightarrow Wb \rightarrow \ell\,\nu b$

escaping detection, and the irreducible electroweak WW production. The $H \rightarrow ZZ \rightarrow 4\ell$ provides instead an extremely clean signature and a reconstructed four-lepton invariant mass with an experimental resolution of 1–2%; however, to achieve a good sensitivity despite the small $Z \rightarrow \ell\ell$ branching fraction, very high efficiency to detect leptons is needed, a challenging task especially at low energies.

As the Higgs boson production cross sections decrease significantly with m_H, to extend the sensitivity at high masses it is necessary to complement the two aforementioned golden modes with searches in other Z and W final states, with larger backgrounds but also larger branching fractions, *i.e.* $H \rightarrow WW \rightarrow \ell\nu\, 2q$, $H \rightarrow ZZ \rightarrow 2\ell\, 2q$, and $H \rightarrow ZZ \rightarrow 2\ell 2\,\nu$.

The information from the searches in all the final states considered is then combined into a single analysis, using the standard model predictions for the different Higgs boson production cross sections and branching ratios, and statistical inference is used to interpret the outcomes in terms of exclusion limits or discovery significances.

The analysis of the data collected by ATLAS and CMS in the 2011 LHC run restricted the mass of a possible standard model Higgs boson to the narrow region $115 \lesssim m_H \lesssim 127\,\mathrm{GeV}/c^2$ at 95% confidence level [19, 20, 21, 22], where a slight excess of events with respect to the expectations from the backgrounds was observed in the both experiments. These first hints of the signal were eventually confirmed by further data collected in the first half of year 2012 at higher energy, combined with improvements in the sensitivities of the anaylses, allowing to establish the observation of a new boson with mass of about $125\,\mathrm{GeV}/c^2$ [23, 24]. First measurements of the properties of this particle are in agreement with the predictions for the standard model Higgs boson, although the presently limited sensitivity of these measurements leaves large room also for beyond-standard-model physics.

Outline of this document

The first chapter of this thesis is devoted to a theoretical introduction to the Higgs boson physics from the perspective of the direct searches at the LHC. In the subsequent chapter, the CMS detector and event reconstruction will be described, again with focus on the aspects that are more relevant to the Higgs boson searches. A slightly more detailed description of track and muon reconstruction will be given, as these are the areas in which I've been more directly involved during my PhD.

The main body of the thesis is on the description of the Higgs boson searches at CMS, with particular emphasis on the three areas to which

I've devoted most of my time the last years, *i.e.* the H $\rightarrow$ WW $\rightarrow$ $2\ell 2\nu$ and H $\rightarrow$ ZZ $\rightarrow$ 4ℓ channels and the combined analyses of all the channels. Due to the very fast pace at which the CMS analyses are being updated and improved, it has not been possible to provide a full and up-to-date description of all of them, and so for some searches the description is based on the analysis of the 2011 data alone, *i.e.* what was available at the time most of this thesis has been written.

ACKNOWLEDGEMENTS

I would like to thank the three people who supervised my work from the start of my PhD to now, Lorenzo Foà, Gigi Rolandi, and Vivek Sharma, for having set me on this research path.

I also wish to thank the many collegues and friends that have accompanied me along these four years at CERN, without which I would have never succeeded in my work, and also those who have provided me invaluable information needed in assembling this thesis. Deliberately in random order, *grazie* a Tommaso, Stefano, Alessandro, Simone, Stefano, Marco, Cristina, Francesco, Steven, Costanza, Alicia, Andrea, Daniele, Luca, Linda, Boris, Nicola, Domenico, Annapaola, Piergiulio, Adish, Michele, Roberto, Giovanni, Clio, Silvia, Monica, Tommaso, Alessandra, Santiago, Lara, Frederic, Nicolò, Virginia, Alessandro, Jean-Roch, Matteo, André, Michalis, Xavier, Jonathan, Si, Volker, Andrea, Guillelmo, Colin, Sara, Wolfgang, Mario, Minshui, Vincenzo, Marco, Luca, Riccardo, Aaron, Zoltan, Roberto, Martina, Markus, Ivan, Roger, Matthew, Josh, Daniele, Giuseppe, Lorenzo, Roberto, Lorenzo, Matthew, Fabian, Clara, Nicola, Andrea, Slava, Giulio, Chiara, Maiko, Silvia, Juerg,...

Chapter 1
The standard model of elementary particles

1.1. Overview

The standard model of particle physics (SM) is a theory describing all known elementary particles and all their known interactions except for the gravitational one, which is anyway irrelevant at microscopic scales.

The SM is formulated as a quantum field theory: matter is described in terms of spin $1/2$ fermions, whose interactions are mediated by spin 1 bosons; finally, there is a scalar Higgs boson field, also interacting with fermions and gauge bosons. The three interactions mediated by the vector bosons are the nuclear strong, electromagnetic and nuclear weak interaction, mediated respectively by gluons, photons, and W and Z bosons.

Three generations of fermions exist, each composed by a pair of quarks with electrical charges $+2/3$, $-1/3$, one lepton of charge -1, and one almost-massless neutrino with no electrical charge; each fermion has an associated anti-fermion with opposite electrical charge.

Quarks. The three doublets of quarks are: up and down, (u, d), stable constituents of the atomic nuclei, with masses of a few MeV/c^2; charm and strange, (c, s), with masses of about 1.3 and $0.1\,\text{GeV}/c^2$; top and bottom (t, b), with masses of about 172 and $4.2\,\text{GeV}/c^2$. Quarks have only been observed in hadrons, bound states of either three quarks (barions), three antiquarks (anti-barions), or one quark and one anti-quark (mesons); an exception is the top quark, whose extremely short lifetime prevents the formation of bound states.

Quark flavour is always conserved by strong and electromagnetic interactions, but not by the weak interactions mediated by W bosons. Decays through weak interactions of hadrons composed of u, d and s quarks are characterized by long lifetimes, with $c\tau$ values ranging from centimeters to meters, while for hadrons containing c and b quarks lifetimes are $O(100\,\mu\text{m})$ and $O(500\,\mu\text{m})$ respectively. Decays mediated by strong or electromagnetic interactions have lifetimes too short to be detectable.

Leptons. The three generations of leptons are the electron e, the muon μ and the tau τ, with masses of about $0.5\,\text{MeV}/c^2$, $0.1\,\text{GeV}/c^2$ and $1.8\,\text{GeV}/c^2$; while only the electron is stable, also the muon can be considered stable in the context of collider experiments due to its long lifetime ($c\tau \sim 660\,\text{m}$). Because of the shorter tau lepton lifetime, $c\tau \simeq 87\,\mu\text{m}$, taus are observed only through their decay products.

Three neutrinos exist, each associated to one of the charged leptons, ν_e, ν_μ, ν_τ. A lepton number is defined in each generation as the number of leptons, plus the number of neutrinos, minus the number of anti-leptons and anti-neutrinos; in the SM, the lepton flavour is conserved separately for each generation, except for neutrino oscillation phenomena which conserve only the total lepton number. Neutrinos interact with other particles only through the nuclear weak force, and are therefore undetectable except in dedicated experiments.

In the contemporary formulation accounting for neutrino masses, usually a heavy neutrino is included in each fermion generation, but these neutrinos play little or no role in the high energy physics phenomenology.

For spin $1/2$ fermions, a chirality can be defined as the projections of the spin along the momentum of the particle, which is a relativistic invariant for massless fermions; in the massless limit, chirality is also preserved in the interactions between fermions and vector bosons.

1.2. The standard model as gauge theory

The framework used to describe the interactions in the SM is that of gauge theories, quantum field theories where particles are endowed with a local symmetry, *i.e.* where the transformation parameters can be function of the space-time position; these symmetries act on some internal degrees of freedom of the particles, and not their space-time degrees of freedom.

A generic gauge theory is formulated by selecting a symmetry group $\mathcal{G}$, arranging all fields in representations of this group, and defining a gauge invariant action $S = \int \mathcal{L}\, d^4x$, which is most often achieved by making the Lagrangian itself invariant.

Fields transform linearly under the action of the gauge symmetry as $\phi'^i(x) = U^{ij}(x)\phi^j(x)$, but space-time derivatives of the fields do not since U^{ij} is dependent on x, so a covariant derivative is introduced as

$$D_\mu \phi^i(x) = \partial_\mu \phi^i(x) - i\, g\, t_a^{ij} A_\mu^a(x)\phi^j(x)\,, \tag{1.1}$$

where $A_\mu^a(x)$ are the gauge vector fields, t_a^{ij} the generators of $\mathcal{G}$ in the representation of the fields ϕ^i and g is the charge[1]. If all ordinary derivatives are replaced with covariant ones then a Lagrangian invariant under the global symmetry is made invariant also under gauge transformations.

The beautiful feature of gauge theories is that the interaction terms in the Lagrangian $\mathcal{L}$ arise naturally from the free Lagrangian just by replacing the ordinary derivatives in the kinetic terms with the covariant ones. This means that the most general gauge theory with vectors and fermions can be synthetically written as

$$\mathcal{L} = -\frac{1}{4} F_{\mu\nu}^a F^{\mu\nu a} + \bar{\psi}(i\slashed{D} - m)\psi \,, \tag{1.2}$$

where the field strength tensor for the gauge bosons $F_{\mu\nu}^a$ is given by

$$F_{\mu\nu}^a = \partial_\mu A_\nu^a - \partial_\nu A_\mu^a - i g f^{abc} A_\mu^b(x) A_\nu^c(x)\,. \tag{1.3}$$

The last term introduces trilinear and quadrilinear terms in the Lagrangian, corresponding to self-interaction between the gauge bosons, except in the case of abelian symmetries, for which there is a single gauge boson and f^{abc} is zero. The symmetry forbids the presence of mass terms for the gauge bosons in the Lagrangian.

As all interactions are determined by the free Lagrangian, gauge theories are very constrained: everything is fixed by the choice of the group, the representation of the fermions (gauge bosons are always in the fundamental representation), and the value of the charges; for non abelian groups (*i.e.* $SU(2)$) there is just one charge, while abelian $U(1)$ can have a different charge for each particle.

1.2.1. The standard model gauge group

In order to describe the strong, electromagnetic and weak interactions of the standard model, the gauge group of the SM is taken to be the direct product $SU(3)_C \times SU(2)_L \times U(1)_Y$.

Strong $SU(3)_C$

$SU(3)$ is the group implementing the strong interactions among quarks and gluons, described by the quantum chromodynamics (QCD). The group has 8 generators, corresponding to the 8 massless gluons, and is non abelian, so the gluon are charged ("coloured"), and have self-interactions.

---

[1] For a simple symmetry group there is just one charge g; otherwise $\mathcal{G}$ can be the product of many groups $\mathcal{G}_i$ each with its own charge g_i.

Quarks fill the simplest non trivial representations of the group, 3 and $\bar{3}$, from the product of which the singlet bilinear $\bar{q}q$ and the octect $\bar{q}\gamma^{\mu}\lambda^{a}q$ are obtained, corresponding to the gauge-invariant mass term, and the covariant vector current coupled to gluons.

Strong coupling and confinement.   At low energies the interactions among quarks are very strong, leading to a complex dynamic that cannot be described as a perturbative expansion around a free theory, although theoretical predictions can still be obtained with other techniques such as lattice calculations. The energy scale at which these effects become important is $\Lambda_{QCD} \sim 250\,\mathrm{MeV}$, which is roughly the mass scale of light hadrons.

It is an experimental fact that all free particles are "colorless" $SU(3)$ singlets: all mesons are in the scalar $\bar{q}^{a}q^{a}$ state, and barions are in the antisymmetrical $\varepsilon^{abc}q^{a}q^{b}q^{c}$. The intuitive explanation for this is that among coloured particles there is an attractive force which increases for increasing distance, but there is yet no well proved and quantitative explanation of this behaviour from a theoretical point of view. In the lack of a full understanding of confinement from first principles, predictions for hadron formation are obtained using phenomenological models tuned on experimental results.

Asymptotic freedom.   In the high energy regime the behaviour of QCD changes: quantum corrections from vacuum polarization cause the interaction to become weaker with decreasing distance, or increasing energy. This behaviour is well understood in the context of renormalization, and is extremely important because it allows a perturbative treatment of the strong interactions.

Electroweak $SU(2)_L \times U(1)_Y$

The electromagnetic and weak interactions emerge from the breaking of a $SU(2) \times U(1)$ symmetry, as explained later in Section 1.3; in this section, the dynamics of these interactions will be described as if the symmetry were not broken.

From the point of view of the electroweak theory, fermions of different chirality are considered different fields, and have different interactions. Because of this, an unbroken $SU(2)_L \times U(1)_Y$ also forbids the presence of fermion mass terms in the Lagrangian.

The two components of the electroweak gauge group are associated to the weak isospin and weak hypercharge.

Weak isospin.   The weak isospin is described in terms of the $SU(2)_L$ group, with three generators corresponding to a triplet of gauge bosons ($\vec{W}$ or W^i). Just like in the $SU(3)$ case, the SM fermions fill in the simplest representations of $SU(2)$: fermions of left-handed chirality are in the doublet $T = \frac{1}{2}$ representation, while right-handed fermions are in the trivial singlet $T = 0$ representation (*i.e.* they do not interact with the $\vec{W}$ bosons).

Weak hypercharge. $U(1)_Y$ is the abelian group associated to the weak hypercharge, corresponding to the interactions mediated by a neutral B boson. In principle, the symmetry would allow for arbitrary values of the charges for each particle, but in the SM constraints among the charges are enforced by the requirements that the symmetry is preserved at quantum level (anomaly cancellation)[2].

The particle content of the SM, with their gauge group representation and hypercharges, including the right handed neutrino, is summarized in Table 1.1. The electrical charge is given by $Q = T_3 + Y$, the sum of the weak hypercharge and the weak isospin ($\pm\frac{1}{2}$ for doublets, 0 for singlets).

		$SU(3)$	$SU(2)$	$U(1)$
	$L = \binom{\nu}{e_L}$	1	2	$-1/2$
leptons	e_R	1	1	-1
(x3 gen.)	N	1	1	0
	$Q = \binom{u_L}{d_L}$	3	2	$+1/6$
quarks	u_R	3	1	$+2/3$
(x3 gen.)	d_R	3	1	$-1/3$
gluons	g	8	1	0
W bosons	$\vec{W}$	1	3	0
B boson	B	1	1	0

Table 1.1. Particle content of the SM from a gauge point of view, except for the Higgs sector. The L and R subscript refer to left-handed and right-handed chiralities; it is omitted for light neutrinos ν (only left-handed) and the possible heavy neutrinos N (only right-handed). The three generations of fermions have exactly the same gauge interactions so they are not stated separately.

1.3. Spontaneous symmetry breaking and the Higgs boson

The SM described in 1.2.1, with only gauge interactions, is far from satisfying. First, the electroweak gauge symmetry requires the fermions and gauge bosons to be massless, which sharply contrasts with the observational evidence. However, an explicit breaking of the symmetry by

---

[2] Constraints between the charges also emerge naturally when the SM is embedded in a larger gauge group with no abelian factors, *e.g.* $SU(5)$, as popular in grand unified theories.

adding extra mass terms in the Lagrangian yields a theory with fundamental problems in the high and low energy limits.

Moreover, gauge interactions are identical for all three generations of particles, *i.e.* that the Lagrangian is invariant under a full $U(3)$ group of unitary rotations in the generation (or flavour) space, so that flavour quantum numbers for quarks in each generation must be separately conserved: this forbids transitions between fermions of different doublets such as those in the observed $K \to \pi\pi$ decays ($s \to du\bar{u}$ at quark level).

1.3.1. Phenomenology of spontaneous symmetry breaking

The spontaneous symmetry breaking is a peculiar feature of infinite dimensional systems like quantum field theories, in which a symmetric Lagrangian can produce a physics which is not symmetric; beyond particle physics, this feature is manifest in a plethora of different environments, from the rotational instability of fluids to condensed matter and solid state theories (superfluids, superconductors). The breaking happens in theories for which the vacuum state of minimum energy is not symmetric: the physical observables are vacuum expectation values of some functions of the fields, $\langle 0|\phi_1(x_1)\ldots\phi_n(x_n)|0\rangle$, which won't be symmetric if the vacuum $|0\rangle$ is not symmetric .

The phenomenology arising from the sponaneous breaking of a gauge symmetry can be introduced first with a minimalistic model: a charged scalar particle represented by the complex field ϕ coupled to a $U(1)$ gauge boson A_μ, for which the Lagrangian is

$$\mathcal{L} = -\frac{1}{4}(F_{\mu\nu})^2 + (D_\mu\phi)^\dagger(D^\mu\phi) + \mu^2\phi^\dagger\phi - \frac{\lambda}{4}(\phi^\dagger\phi)^2 \quad (D_\mu = \partial_\mu - i\,e\,A_\mu).$$
(1.4)

The potential depends only on $\rho^2 = \phi^\dagger\phi$, $V(\rho^2) = -\mu\rho^2 + \frac{\lambda}{4}\rho^4$, and for $\mu > 0$ it has minimum on the circumference $\rho^2 = \frac{2\mu}{\lambda}$, not in the origin $\phi = 0$: this means that potential will be minimized for a vacuum expectation value the field $v = \langle\phi\rangle$ different from zero. The perturbative expansion of the theory is done using $\varphi = \phi - v$ as dynamical field, and v can be taken to be along the real axis. After the reparametrization $\varphi = \rho e^{i\theta}$ to express ϕ in terms of two real fields ρ, θ, the covariant derivative of ϕ becomes

$$D_\mu\phi = \left[\partial_\mu\rho + i\,\rho\,(\partial_\mu\theta - e\,A_\mu) - i\,e\,A_\mu(\rho - v)\right]e^{i\theta}.$$
(1.5)

Under a gauge transformation of parameter Λ, fields transform as $\theta \to \theta + e\Lambda$ and $A_\mu \to A_\mu - \partial_\mu\Lambda$, so an appropriate choice of $\Lambda(x)$ can rotate away θ from the Lagrangian completely leading to

$$\mathcal{L} = -\frac{1}{4}(F_{\mu\nu})^2 + (\partial_\mu\rho)^2 + e^2(\rho - v)^2 A_\mu A^\mu + V(\rho). \qquad (1.6)$$

This Lagrangian appears much different from the original one: there is a single, neutral scalar particle (ρ), interacting with the gauge boson through trilinear and quadrilinear vertices ρA^2 and $\rho^2 A^2$, and a mass term for the gauge boson $e^2 v^2 A_\mu A^\mu$.

1.3.2. Electroweak symmetry breaking in the standard model

In the standard model the Higgs field before symmetry breaking is a $SU(2)_L$ doublet Φ with weak hypercharge $\frac{1}{2}$; its covariant derivative can be written as

$$D_\mu \Phi = \partial_\mu \Phi - \frac{1}{2} i\, g\, \vec{W}_\mu \cdot \vec{\sigma}\, \Phi - \frac{1}{2} i\, g'\, B_\mu \Phi. \qquad (1.7)$$

In the parametrization where the upper and lower components of Φ are separately denoted as ϕ^+ and ϕ (both charged under $U(1)_Y$), and working in the unitary gauge for which $\langle\Phi\rangle$ is real, then the expansion of the field around the vacuum state of the broken symmetry is

$$\Phi = \begin{pmatrix} \phi^+ \\ \phi \end{pmatrix} \xrightarrow{SSB} \begin{pmatrix} 0 \\ v \end{pmatrix} + \begin{pmatrix} \xi^+ \\ \xi \end{pmatrix}. \qquad (1.8)$$

Except for a more complex algebra, the conclusions are similar to those of the simplified $U(1)$ model: the terms quadratic in v will provide mass to the gauge bosons through terms

$$\mathcal{L} \supset \frac{1}{8}v^2 \left(g^2 |W^1|^2 + g^2 |W^2|^2 + (-g W_\mu^3 + g' B_\mu)^2 \right) \qquad (1.9)$$

which can be diagonalized to obtain four mass eigenstates

$$W_\mu^\pm = \frac{1}{\sqrt{2}}(W_\mu^1 \mp i\, W_\mu^2) \qquad m_W = g\frac{v}{2}$$

$$Z_\mu^0 = \frac{1}{\sqrt{g^2 + g'^2}}(g\, W_\mu^3 - g'\, B_\mu) \qquad m_Z = \sqrt{g^2 + g'^2}\,\frac{v}{2}$$

$$A_\mu = \frac{1}{\sqrt{g^2 + g'^2}}(g\, W_\mu^3 + g'\, B_\mu) \qquad m_A = 0$$

The first two fields describe the W and Z bosons mediating weak interactions, and the last one the photon. Of the four components of the Φ field, only one is left after the gauge transformation, the neutral scalar H referred to as the Higgs boson.

Fermion masses and flavour quantum numbers. The same Higgs field that gives masses to the gauge bosons can also provide masses to the fermions: gauge invariance permits the following Yukawa terms to appear in the SM Lagrangian

$$\mathcal{L} \supset \lambda_u \bar{Q} \cdot \Phi^\dagger u_R + \lambda_u \bar{Q} \cdot \Phi d_R + \lambda_e \bar{L} \cdot \Phi e_R + \lambda_\nu \bar{L} \cdot \Phi^\dagger N , \quad (1.10)$$

where the dot product is the scalar product in $SU(2)$ space.

The Yukawa couplings λ_x, 3×3 matrices because of three generations, result in mass matrices $m = v\lambda$ when Φ acquires a vacuum expectation value; the last term provide neutrino masses, but otherwise plays a negligible role in SM physics. In the case of leptons, when neglecting neutrino masses it is possible to redefine the fields so that the three lepton generations are simultaneously eigenstates of the mass matrix and of the $SU(2)$ interactions. This is not possible in the case of quarks, since there is one extra matrix to diagonalize; it is conventional to define the quark generations in the mass eigenstate basis, leaving a unitary mixing matrix V_{CKM} in the interaction part of the Lagrangian.

The introduction of Yukawa couplings also drastically reduces the global symmetry of the theory, as generations are no longer identical. With no neutrino masses, the total symmetry is $U(1)^4$: one from the global phase of all quarks, barion number conservation, and one for each lepton flavour (which are separately conserved). If neutrino masses are added, but with no Majorana mass for N, it is only $U(1)_B \times U(1)_L$, and the total lepton and barion number (L, B) are conserved; in the most general case another phase is constrained by the additional $m_N \bar{N}^c N$ term, and L is no longer conserved.

1.4. Standard model Higgs boson production mechanisms

In the standard model, Higgs boson production in proton-proton collisions can happen through four main modes: gluon fusion gg $\rightarrow$ H; vector boson fusion qq $\rightarrow$ H + 2 jets, associated production of a Higgs boson with a W or Z boson, and associated production with a t$\bar{\text{t}}$ pair. The hierarchy of the cross sections and their dependence on the Higgs boson mass is shown in Figure 1.1; in general all production cross sections decrease with increasing Higgs boson mass, except for gg $\rightarrow$ H around the $m_H \sim 2m_t$ threshold. A comprehensive collection and review of the most accurate theoretical predictions for the Higgs boson production cross section in proton-proton collisions at $\sqrt{s} = 7$ TeV is available in reference [10].

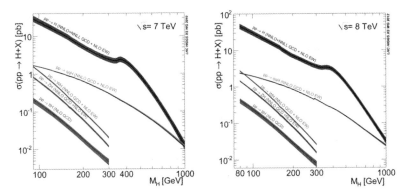

Figure 1.1. Theoretical predictions for the Higgs boson production cross sections in proton-proton collisions at $\sqrt{s} = 7\,\text{TeV}$ (left) and $8\,\text{TeV}$ (right). The five production cross sections, in decreasing order, are gluon fusion, vector boson fusion, WH, ZH and $t\bar{t}$H [10].

Gluon fusion. The main Higgs boson production mechanism at the LHC is through gluon fusion, for which the leading Feynman diagram involves a quark loop (Figure 1.2). A remarkable property of the amplitude for this process at leading order is that it is zero for massless quarks, and saturates to a constant for increasing m_q/m_H, so that the cross section is proportional to the square of the number of heavy quarks; in the standard model, the amplitude is dominated by the contribution from the top quark, the only one with mass scale comparable to the Higgs boson mass, but the production cross section is naturally sensitive to contributions from hypothetical particles with QCD interactions beyond the standard model ones at higher energy scales.

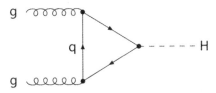

Figure 1.2. Leading diagram for gg → H production.

Higher order QCD corrections to the process are positive and large, increasing the cross section by about 100% at the next-to-leading order (NLO), 25% at next-to-next-to-leading order (NNLO), and a further 8% when leading logarithmic corrections are resummed (NNLL); the state of the art theoretical predictions also include NLO corrections from electroweak and mixed QCD-electroweak terms, whose size is nonetheless small.

Because of the slow convergence of the QCD radiative corrections, the theoretical uncertainty on the predicted cross section from neglected higher order terms is 8%–12% depending on the Higgs boson mass even for the most precise NNLO+NNLL result. The theoretical uncertainties from the knowledge of the gluon parton density function are likewise sizable, about 8% up to Higgs boson masses below $250\,\mathrm{GeV}/c^2$ and increasing up to 10% for $m_\mathrm{H} = 600\,\mathrm{GeV}/c^2$.

As a natural consequence of the large QCD corrections, final states with one or more hadronic jets are not uncommon, introducing complications in the analyses performed in exclusive jet multiplicity final states, e.g. $\mathrm{H} \to \mathrm{WW} \to 2\ell 2\,\nu$.

Vector boson fusion. With a production cross section of about one tenth of the gluon fusion one, vector boson fusion is the second largest production mode relevant at the LHC. The three leading order diagrams for the process in the t, u and s channels are shown in Figure 1.3; in the first two, either or both of the incoming quarks can also be replaced by anti-quarks, but the qq initial state gives the largest contribution for proton-proton collisions. From an experimental point of view, the t and

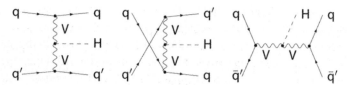

Figure 1.3. Leading order Feynman diagrams for qq $\to$ H $+$ 2 jets production through vector boson fusion, in the channel from left to right.

u channels are the most relevant ones, because the momentum transfer between the two scattering quarks tends to be of order m_V, small compared to $\sqrt{s}$, resulting in final states with two jets at high rapidities. As the two outgoing quarks are not connected by any quark or gluon line, QCD radiation in the region between the two jets is suppressed, another experimentally viable feature, useful e.g. in separating a qqH $\to$ qqWW $\to 2\ell 2\,\nu + 2$ jets signal from the larger $t\bar{t} \to b\bar{b}$WW $\to 2\ell 2\,\nu + 2$ jets background.

For this process, QCD radiative corrections are significantly smaller, $\mathcal{O}(10\%)$ at NLO. As a consequence, the theoretical uncertainty from higher order terms on the most accurate predicion available, at NNLO accuracy for QCD effects and NLO for electroweak ones, is also smaller: below 0.5% for Higgs boson masses up to $200\,\mathrm{GeV}/c^2$, and below 2% up to $600\,\mathrm{GeV}/c^2$. As the parton density function for quarks is also known to higher accuracy than the corresponding one for gluons, the theoretical uncertaintes from this source are also smaller, ranging from about 2.5% at low m_H to about 7.5% at 600 GeV/c^2.

Associated production. The associated production of a Higgs boson and a W or Z boson is characterized by an even smaller production cross section than vector boson fusion. This mode is nonetheless experimentally viable when considering Higgs boson decays to bottom quark pairs, since leptons and neutrinos from the W or Z decay can provide handles to select events, while the all-hadronic final states from gg $\rightarrow$ H $\rightarrow$ b$\bar{\text{b}}$, qq $\rightarrow$ qqH $\rightarrow$ qqb$\bar{\text{b}}$ suffer larger backgrounds from QCD multijet production.

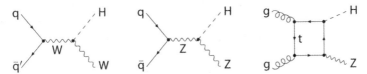

Figure 1.4. Leading order diagrams for WH and ZH associated production, and higher-order contribution to ZH with gluons in the initial state.

The main Feynman diagrams for this process are closely related to the vector boson fusion ones since they both rely on the WWH and ZZH vertices (Figure 1.4). As for vector boson fusion, the cross section has been computed at NNLO for QCD and NLO for electoweak, yielding an overall uncertainty from higher orders of 1% for WH in the explored mass range 110–200 GeV/c^2. The uncertainty for ZH is about twice that for WH, due to the presence of NNLO diagrams with gluons in the initial state, whose contribution to the total cross section is small, 2%–6% depending on m_H, but has a large relative uncertainty from QCD radiation similarly to gg $\rightarrow$ H. Uncertainties from the knowledge of the parton density functions are in the 3%–4% range for both processes.

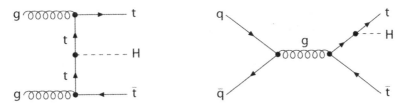

Figure 1.5. Leading order diagrams for t$\bar{\text{t}}$H associated production with gluons and quarks in the initial state.

Associated t$\bar{\text{t}}$H (Figure 1.5) is an important channel, allowing for a direct measurement of the Yukawa coupling between top quark and Higgs boson, but the very low production cross section and the complex final state with a large number of hadronic jets make it very challenging to reconstruct properly. The cross section has been computed to NLO in QCD only, and the corresponding estimate of the uncertainty from unknown higher order terms is about 10%. As the largest contribution to the production cross section is from the partonic initial state with two gluons, the uncertainty from the knowledge of the parton density functions is similar to that for gluon fusion, in the 8–10% range.

1.5. Standard model Higgs boson decays

The standard model Higgs boson is an unstable particle with immeasurably short lifetime, so it can be detected only through its decay products. As for any unstable particle, the branching fractions are determined by the partial widths of the decays into each final state,

$$\mathrm{BR}(H \to X) = \frac{\Gamma(H \to X)}{\sum_Y \Gamma(H \to Y)} \quad (1.11)$$

where each partial width depends only on the square of the couplings of the Higgs boson to those specific decay products and on kinematic factors.

At leading order in the SM couplings, the Higgs boson can decay into pairs of heavy fermions through Yukawa interactions (Figure 1.6, left), and into pairs of W or Z bosons through $SU(2)_L$ interactions (Figure 1.6, right). The Yukawa decays have partial widths proportional the square

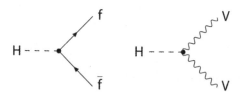

Figure 1.6. Leading order Feynman diagrams for Higgs boson decays into pairs of fermions (left) and gauge bosons (right).

of the ratios between fermion mass and Higgs boson vacuum expectation value $(m_f/v)^2$, so for m_H below the $t\bar{t}$ threshold the decays are preferentially to $b\bar{b}$ pairs, with smaller contributions from $c\bar{c}$ pairs, $\tau^+\tau^-$ pairs, and negligible ones from the other fermion pairs. The decays into gauge bosons are dominant above the $m_H = 2m_V$, but can happen also for lighter Higgs bosons decaying into a pair of off-shell W and Z boson which further decay into four fermions. While the partial width of WW and ZZ decays is strongly suppressed when the bosons are off shell, these decays can become competitive to the $H \to b\bar{b}$ one already below the threshold since the latter is suppressed by $(m_b/v)^2$. Interference effects in four-fermion decays are important at low mass, *e.g.* $O(10\%)$ at $m_H = 120\,\text{GeV}/c^2$, while they are below 1% above threshold.

Loop-induced decays into pairs of photons and gluons are also important in the low mass region. The decay into gluons is the mirror process of the gluon fusion production, happening through a quark loop (Figure 1.2); the large Yukawa coupling of the top quark and the eightfold color multiplicity of the final state allow this loop decay to be competitive to the tree-level ones into $c\bar{c}$ pairs, $\tau^+\tau^-$ pairs.

Higgs boson decays into diphotons can happen both through a fermion loop, dominated by the top quark contribution, and through a W boson loop with two $WW\gamma$ vertices (Figure 1.7). The partial width of the $H \rightarrow \gamma\gamma$ decay is about a factor 40 smaller than the one for $H \rightarrow gg$, due to the smaller electroweak couplings, the reduced degeneracy of the final state and the destructive interference between the two processes. Decays to $Z\gamma$ are also possible through similar loops, but they are less relevant experimentally because $H \rightarrow Z\gamma \rightarrow \ell\ell\gamma$ final states are further suppressed by the small $BR(Z \rightarrow \ell\ell)$, and the other Z decays are less accessible experimentally.

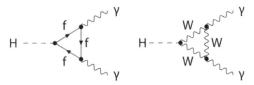

Figure 1.7. Leading order Feynman diagrams for Higgs boson decays into pairs of photons.

The dependency of the branching fractions on the Higgs boson mass is shown in Figure 1.8. Theoretical uncertainties on the partial widths are due to the missing higher orders in the perturbation theory and from the knowledge of the strong coupling constant and quark masses; typical scales of these uncertainties are; 0.5% for VV, 1% for $\gamma\gamma$, 2% for $\tau\tau$ and $O(5\%)$ for the other modes. Due to the interplay of the different partial widths in the definition of the branching fractions through equation (1.11), the uncertainties from all decay widths contribute to each branching fraction as

$$\frac{\delta BR(H \rightarrow X)}{BR_{H \rightarrow X}} = (1 - BR_{H \rightarrow X}) \frac{\delta\Gamma_{H \rightarrow X}}{\Gamma_{H \rightarrow X}}$$
$$+ \sum_{Y \neq X} BR_{H \rightarrow Y} \frac{\delta\Gamma_{H \rightarrow Y}}{\Gamma_{H \rightarrow Y}} . \tag{1.12}$$

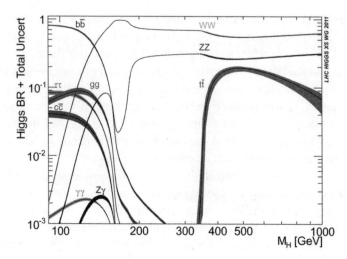

Figure 1.8. Predicted branching fractions for the standard model Higgs boson decays, as function of the Higgs boson mass. The theoretical uncertainties on the predictions is displayed as a band [11].

The main consequences of equation (1.12) are a suppression of the uncertainty on $BR(H \to b\bar{b})$ at low mass, and an increase of the uncertainty on $BR(H \to VV)$ above the $t\bar{t}$ threshold.

Chapter 2
The CMS experiment at the CERN LHC

2.1. The Large Hadron Collider

The Large Hadron Collider (LHC) is a superconducting circular proton-proton collider operating at the laboratories of the European Organization for Nuclear Research (CERN); it is hosted in the 27 km underground tunnel that was previously used for the Large Electron Positron Collider (LEP). The design specifications of the LHC targeted operations at a center-of-mass energy of 14 TeV, with an instantaneous luminosity of $2 \cdot 10^{33}$ cm^{-2}s^{-1} in the first few years and $1 \cdot 10^{34}$ cm^{-2}s^{-1} afterwards [16].

The startup strategy was revised afterwards, selecting lower energies for the first years of running, 7 and 8 TeV, motivated by the reduced time needed for machine commissioning and the safer operational regime for the superconducting magnets. Operation at an energy close to the design one is foreseen for the end of 2014.

Physics processes and event rates at the LHC. In proton-proton collisions, the production rates for different physical processes depend strongly on the energy scale of the process, and not just on the characteristic strength of the interactions, as shown in Figure 2.1. In order to reach sensitivity to rare processes like the production of a SM Higgs boson instantaneous luminosities of order 10^{33} cm^{-2}s^{-1} are needed, resulting in a very large rate for more abundant or lower energy processes.

Low energy, inelastic QCD interactions are the dominant process at the LHC, with event rates $O(10^8)$ Hz for an instantaneous luminosity of 10^{33} cm^{-2}s^{-1}, but are easily suppressed by the requirements of a significant flux of particles not collinear to the beam line, or the presence of leptons or photons. More challenging backgrounds are given by the production of heavy flavour quarks, which can decay into leptons, and by higher energy QCD multi-jet events, with event rates $O(10^5)$ Hz. These event rates can be compared to the maximum rate at which the data from all the channels of a general-purpose LHC detector like CMS can be read out, about 10^5 Hz, and the maximum rate at which events can be stored persistently for physics analysis, $O(100)$ Hz; a very fast event selection, relying initially only on a subset of the detector channels, is therefore an essential requirement for any LHC experiment.

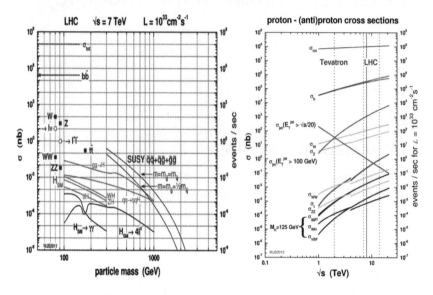

Figure 2.1. Production cross sections for different physical processes at the LHC, and corresponding event rates at an instantaneous luminosity of 10^{33} cm^{-2}s^{-1}. Cross sections are shown as function of the energy scale of the process for a fixed collision energy $\sqrt{s} = 7$ TeV in the left plot, and as function of $\sqrt{s}$ in the right plot [25].

Operations.   The collider became operational in November 2009 at injection energy, 450 GeV per beam, and soon afterwards at an energy of 1.18 TeV per beam. Operations at 3.5 TeV per beam started in spring 2010, with a rapidly increasing luminosity profile, allowing CMS to collect about 0.3 pb^{-1} of integrated luminosity usable for analysis by summer 2010, 36 pb^{-1} by end of 2010, and 4.8 fb^{-1} by end of 2011. In the last months of 2011, instantaneous luminosities up to $3.5 \cdot 10^{33}$ cm^{-2}s^{-1} have been achieved (Figure 2.2). In 2012, the beam energy has been increased to 4 TeV per beam, with instantaneous luminosities close to $6.5 \cdot 10^{33}$ cm^{-2}s^{-1}, allowing the accelerator to deliver about 6 fb^{-1} of integrated luminosity to the CMS and ATLAS experiments by the summer.

In 2011 and 2012, the LHC was operated with a bunch spacing of 50 ns, and a corresponding maximum capacity of about 1400 colliding bunches; as the cross section for inelastic pp collisions at $\sqrt{s} = 7$ TeV is approximately 70 mb, the expected number of interactions per bunch crossing is 5–18 for instantaneous luminosities of 1–$3.5 \cdot 10^{33}$ cm^{-2}s^{-1}.

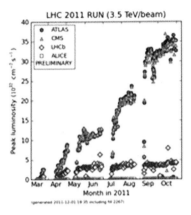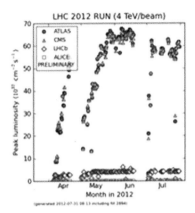

Figure 2.2. LHC peak instantaneous luminosity during the 2011 run at $\sqrt{s} = 7\,\mathrm{TeV}$ (left) and the 2012 run at $\sqrt{s} = 8\,\mathrm{TeV}$ (right).

Similarly, the 2012 running has been characterized by approximately twice that number of interactions per bunch crossing. This pile-up of multiple interactions poses significant challenges to the trigger, event reconstruction and analysis, as it increases significantly the complexity of the events[1].

For detector components that integrate signals over a time window larger than 50 ns, an additional challenge is posed by the contaminations from particles produced in previous or subsequent collisions (out-of-time pile-up).

2.2. Overview of the CMS experiment

The Compact Muon Solenoid experiment is one of the two general-purpose experiments designed to search for new physics at the new energy and luminosity frontiers that the LHC is opening. The discovery or exclusion of the SM Higgs boson, and more generally the understanding of the electroweak symmetry breaking mechanism, has been one of the primary physics goals of CMS already since its design phase.

Detector. The Compact Muon Solenoid detector has a roughly cylindrical shape, 22 m long and with an outer radius of 7 m. It is divided into two regions: barrel ($|\eta| \leq 1.2$), where subdetectors are layered at increasing values of the radius R, and endcaps ($|\eta| > 1.2$) where layers

---

[1] 15 inelastic pp collisions yield a total flux of particles similar to a $t\bar{t}$ event.

are arranged along the z coordinate[2].

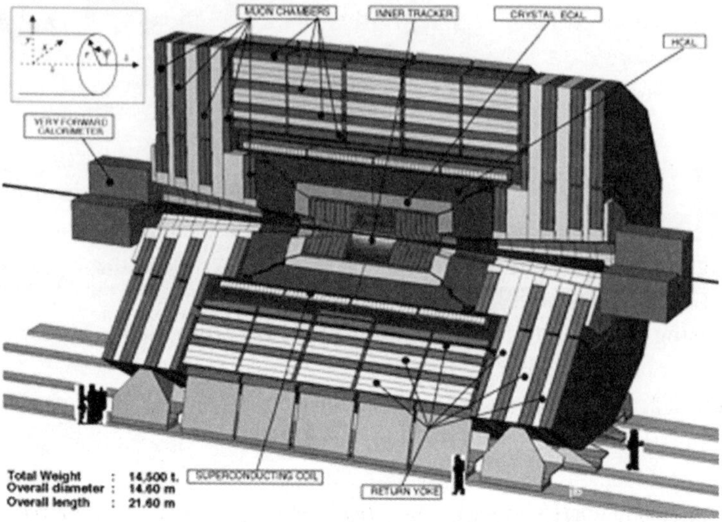

Figure 2.3. Longitudinal view of the CMS detector. The cartesian and cylindrical coordinates are defined as shown in the top right inset; the x axis points towards the center of the LHC ring.

The inner tracker, composed of silicon pixel and silicon strip sensors, is located at the center of the detector, and immersed in the uniform magnetic field of 3.8 T produced by a superconducting solenoid. The coverage of the inner tracker extends up to $|\eta| = 2.5$.

The electromagnetic calorimeter (ECAL), composed by lead-tungstenate crystals, and the hadronic calorimeter (HCAL), made of layers of brass and scintillator plates, are located between the tracker and the solenoid. The coverage of ECAL and HCAL extends up to $|\eta| = 3$, and is complemented in the region $3 < |\eta| < 5$ by a forward calorimeter (HF), made of quartz and scintillating fibers.

The muon spectrometer is located in the steel return yoke of the magnet, and uses three technologies of gas detectors: drift tubes (DT), cathode strip chambers (CSC) and resistive plate chambers (RPC), with a coverage up to $|\eta| = 2.4$.

A full description of the CMS detector can be found in reference [17].

Trigger. CMS uses a two-level trigger strategy: the first level (L1), uses information from the calorimeters and muon detectors to select, in

---

[2] CMS uses a cylindrical coordinate system with the z axis along by the beam line. The pseudorapidity η is defined as $\eta = -\ln[\tan(\theta/2)]$, with θ being the polar angle.

less than 1 μs, the most interesting events. It is composed of custom hardware processors, and has an output event rate of about 100 kHz. The high level trigger (HLT) further decreases the event rate to about 300 Hz before data storage. To ensure maximum flexibility, the HLT is implemented as a processor farm running the same software framework used for offline reconstruction.

The full detector readout is available at HLT, but in order to meet the timing requirements given by the input rate from L1, events are discarded before being fully reconstructed, as soon as the available information is enough to take the decision. This allows, for example, the usage of the full track reconstruction only for the events which can't be rejected before using information from the calorimeters or from the faster pixel-only track reconstruction.

Particle Flow event reconstruction.   A distinctive feature of the CMS reconstruction is the use of a particle flow event description, in which each single particle is reconstructed and identified using an optimized combination of all subdetector information. The tasks whose performance benefits most from this technique are jet and missing transverse energy reconstruction and tau identification.

2.3. Track and vertex reconstruction

The CMS inner tracker system is used to detect charged particles and measure their momentum through the bending in the magnetic field. There are several aspects of the searches for the Higgs boson that depend crucially on the performance of the inner detector and tracking.

Most Higgs analyses rely on leptons, for which information from the inner detector is used to determine the momentum, together with information from ECAL for electrons and with information from the muon system for high p_T muons. The inner detector is also used to determine the impact parameter of the leptons, to impose the requirement that all leptons are promptly produced[3] and come from the same pp interaction. Detailed information from the tracker is also used to reject muons from $\pi^{\pm}/K^{\pm}$ decays and electrons from photon conversions, as it will be described in Sections 2.4.3 and 2.5.4.

The reconstruction of charged hadrons is fundamental for the particle flow event reconstruction, since charged hadrons constitute two-thirds of the rate of particles produced in pp collisions. Reconstructed hadron tracks are also very useful to select isolated leptons or photons from

---

[3] This request is slightly relaxed in the searches using taus decaying leptonically, due to the finite lifetime of the tau.

signal processes rejecting the backgrounds from heavy quark decays or misidentified particles, normally associated with hadronic jets; isolation determined from charged hadrons is also naturally robust against pile-up, since hadrons whose tracks are not associated to the signal pp interaction can be removed from the computation.

Efficient charged hadron reconstruction, precise impact parameter measurements and secondary vertex identification are also the main ingredients for the algorithms used to tag jets from heavy flavour decays. Beyond the obvious applications in the H $\rightarrow$ b$\bar{\text{b}}$ searches, heavy flavour tagging is also used to identify Z $\rightarrow$ b$\bar{\text{b}}$ decays in H $\rightarrow$ ZZ searches, and to reject backgrounds from t$\bar{\text{t}}$ and single top processes in the searches in several other final states, $e.g.$ H $\rightarrow$ WW $\rightarrow$ $2\ell2$ ν.

2.3.1. Inner detector layout

Closest to the interaction point tracks are measured with silicon pixel detectors, arranged in three barrel layers and two endcap disks; pixels provide true 2d points with 10–15 μm resolution, have very low occupancy and a signal over noise ratio of ~ 70.

At a distance from the IP where the flux of particles is sustainable, silicon strip detectors are used, arranged in 10 barrel layers and 12 endcap disks. In the barrel strips are parallel to the z axis, while in the endcaps they are along the radial coordinate ρ, to provide in both cases a precise measurement of the $r\phi$ coordinate ($\sigma \sim$ 20–50 μm).

In order to measure also the z coordinate with a precision better than the strip length (about 10 cm), some tracker layers contain an additional set of sensors tilted by 100 mrad with respect to z. Matching the hits on the $r\phi$ detectors to those on the tilted detectors, it is possible to reconstruct 2d points with a z resolution of 200–500 μm.

A schematic cross section of the inner detector is shown in Figure 2.4.

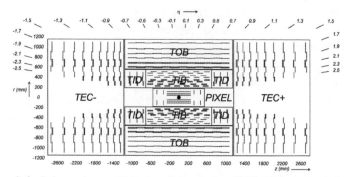

Figure 2.4. Schematic cross section through the CMS tracker. Each line represents a detector module. Double lines indicate back-to-back modules which deliver stereo hit.s

2.3.2. Track reconstruction algorithm

Tracks in the inner detector are normally reconstructed using multiple iterations of the Combinatorial Kalman Filter (CKF) algorithm [26, 27], each composed by three steps: seeding, pattern recognition and final fit. After each iteration, tracks are passed through a quality filter to reject fakes, and hits corresponding to good tracks are removed from the input lists of the subsequent iterations to simplify the pattern recognition. Finally, track lists from the different iterations are merged.

Seeding.   The first step of track reconstruction is to search for pairs or triplets of hits compatibles with the hypothesis of a track coming from the interaction region, and estimating from them the helix parameters. The beam spot position in the transverse plane is used in the seeding steps based on hit pairs, as the information from two hits is not enough to determine the kinematics of the particle; this is a good approximation for primary particles, since the transverse size of the beam spot is very narrow.

Two techniques are used to reduce the combinatorics of seed finding, allowing the algorithms to be affordable also at large pile-up: first, hits whose charge distribution in the strip or pixel detector is manifestly incompatible with the incidence angle of the track can be ignored; second, information from pixel-only vertex reconstruction[4] can be used to reduce the combinatorics of hit pairs and improve the estimates of the helix parameters in the ρ-z plane.

The first iterations of tracking use seeds made from hits in the pixel detector. Subsequent iterations rely on the strip tracker, using matched hits from the pairs of back-to-back modules, to cover the projective gaps in the pixel detector geometry and to allow the reconstruction of detached tracks, *e.g.* from photon conversions or nuclear interactions in the detector material or K_S and Λ decays.

CKF pattern recognition.   Seeds with an initial estimate of the trajectory parameters are propagated layer by layer through the detector looking for compatible hits and updating the helix parameters with the information from the subsequent layers using the Kalman Filter algebra. In case multiple compatible hits are found when extrapolating the helix to a single layer, the algorithm will create one trajectory candidate for each hit and they will be propagated independently.

---

[4] Pixel vertices are reconstructed using simple tracks built from triplets of hits in the pixel detector; the selection is loose enough that also primary vertices from minimum bias interactions can be reconstructed with high efficiency.

Trajectory candidates are also allowed to cross a layer without leaving hits, to account for detector inefficiencies, but the building of candidates is interrupted if they collect too many missing layers, which is usually signature of fake tracks coming from randomly aligned hits. The reconstruction algorithm is interfaced to the detector condition database, so that tracks are not penalized for missing hits associated to detector channels known to be faulty. A similar policy is applied on an event-by-event basis for channels whose raw data is corrupted or contains error flags from the front end electronics.

The main pattern recognition is performed inside-out, but once the track is completed another search is performed backward starting from the outermost hit of the seed: the pattern recognition has a higher hit collection efficiency than the seeding, because it handles correctly the geometrical overlaps between modules on the same layer, and it can also recover hits in pixel layers even when the track was seeded from si-strip hits. This hit recovery procedure is useful to improve the accuracy of the impact parameter measurement, for heavy flavour tagging or rejection of secondary leptons.

After all the candidates have been constructed, a cleaning procedure is applied after each iteration by looking for candidates that share a large fraction of hits, and selecting only the best among them according to the number of hits and the χ^2 of the fit.

Track fitting.   Once all the hits of a track candidate are collected, the Kalman Filter fit is applied again to determine the most accurate estimate of the helix parameters at each point, and to remove any bias introduced at the seeding step (e.g from using the beam spot position in case of hit pair seeding). At this step, outlier hits that were collected during pattern recognition but are eventually found to be incompatible with the overall fit of the trajectory are rejected; this results in a better distribution of the χ^2 for good tracks, and therefore a better discrimination against combinatorial fakes, and allows to reduce the tails of the impact parameter resolution.

Track quality filters.   In order to keep a high efficiency, large search windows and loose requirements are used during pattern recognition, which results in a non negligible fake rate (up to some 10%). To reduce the fake rate, a selection is performed afterwards by applying requirements on the normalized χ^2 of the track, and for the iterations targeting primary tracks also on the longitudinal distance from the closest pixel-only vertex, and on transverse impact parameter and their uncertainties; the selection is dependent on the number of reconstructed hits in the track, as tracks with more hits have a lower fake rate, which allows for looser thresholds. Different quality assignments are possible for every track, corresponding to different working points in the efficiency/fake rate plane.

Hit removal. Tracks passing the tightest quality filter from each step are used to remove hits from the inputs of the next step. A tight requirement on the compatibility χ^2 between track and hit is used, to reduce the impact of mis-assigned hits on the efficiency of the subsequent steps.

In dense environments, charge deposits from close-by particles can give rise to a single reconstructed hit. A requirement on the number of strips or pixels in the cluster can be used to avoid removing these hits in the iterative tracking, so that the particles can be reconstructed even in different iterations.

Track list merging. Once all steps of iterative tracking have been performed, the output collections of the different steps are merged. As the hits from low quality tracks are not removed in iterative tracking, the same particle can be reconstructed by multiple steps; this duplication is removed at the merging step by selecting only the best track among those that share a large fraction of hits.

2.3.3. Track reconstruction efficiency and fake rate

The performance of the track reconstruction algorithm on simulated $t\bar{t}$ events is shown in Figure 2.5 for two pile-up configurations: no pile-up, and an average pile-up of eight interactions per bunch crossing. The efficiency is defined as the fraction of prompt charged particles from the signal vertex that are successfully reconstructed, while the fake rate is defined as the fraction of reconstructed tracks that do not correspond to any simulated particle. A high reconstruction efficiency for charged hadrons below 1 GeV is very useful in the context of jet reconstruction, since those particle carry a sizable fraction of the jet energy and do not reach the calorimeters because of the bending in the magnetic field.

The main source of inefficiency in the reconstruction of charged hadrons are nuclear interactions in the detector material, that result in shorter tracks which might not satisfy the quality requirements applied to reject fakes. Muons are not affected by this issue, and therefore can be reconstructed with an efficiency well above 99% within the full coverage of the muon system, allowing to preserve a very high signal efficiency even in signal final states with many muons, *e.g.* H $\rightarrow$ ZZ $\rightarrow$ 4 μ. Electrons are characterized by larger and non-Gaussian energy losses in the material, and are therefore reconstructed using a dedicated algorithm.

The determination of the tracking efficiency for charged hadrons in data is particularly challenging, as it is difficult to infer the existence of a charged particle in the tracker without a complementary detector to

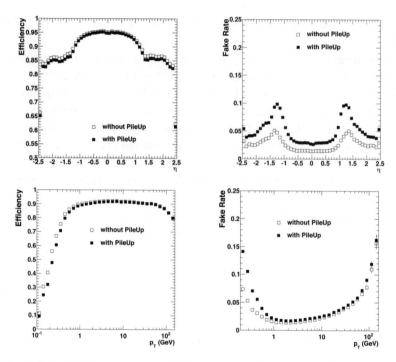

Figure 2.5. Track reconstruction efficiency for prompt tracks in simulated t̄t events, and corresponding fake rate. The two scenarios considered are no pile-up and an average of eight pile-up interactions per bunch crossing. The increase in the fake rate for $|\eta| \sim 1.4$ region corresponds to the transition between the barrel and the endcaps, where larger tolerances are used in the reconstruction to cope with the more difficult geometrical extrapolations and the multiple scattering in the barrel support structures and services (cooling, data and power lines).

identify it. Indirect measurements have been performed by comparing the rates of reconstructed $D^0 \rightarrow K^\pm \pi^\mp$ and $D^0 \rightarrow K^\pm \pi^\mp \pi^+ \pi^-$ decays, for which the relative branching fractions are known; the inferred ratio of tracking efficiencies in data and simulation was found to be close to unity, with an uncertainty of 3.9%. Another estimate was done by embedding simulated tracks in data events, and likewise resulted in an efficiency compatible with the expectation from simulations within 1%. The two measurements are based on early 2010 data, and are reported in reference [28].

The efficiency for muons is measured using the *tag-and-probe* method: dimuons from Z or J/Ψ decays are selected using only information from the muon subdetector for one of the two muons; then, a search is performed in the inner detector for a track compatible with the muon kinematics inferred from the muon subdetector; the tracking efficiency is given by the fraction of muons for which the search is successful. At low momentum, the measurement is made more challenging by the uncertainties in the extrapolation of the track through the detector material between the muon system and the interaction region, and by the large multiplicity of charged hadrons which could result in spurious matches. Measurements have been done on early 2010 data using J/Ψ decays [28], and on the full 2010 sample using Z decays [29], in both cases obtaining an agreement between data and simulations at the 1–2% level. This agreement is confirmed by the results on the full 2011 dataset, as shown in the top panels of Figure 2.6. In 2012 data, where tighter requirements have been applied in several steps of the track reconstruction algorithm to cope with the larger pile-up multiplicity within the same computing

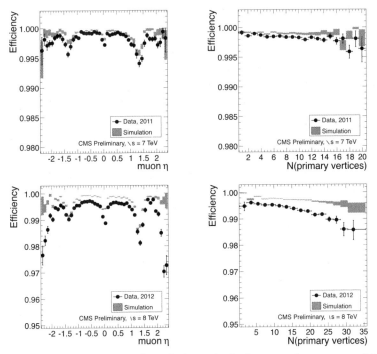

Figure 2.6. Muon reconstruction efficiency in the inner tracker, measured from Z decays using the tag-and-probe method, as function of the muon pseudorapidity (left) and of the number of reconstructed primary vertices (right), for 2011 data (top) and 2012 data (bottom). The efficiency from data is compared to the result of the same procedure applied to simulated events. Small drops in efficiency are visible in the transition between barrel and endcap regions ($|\eta| \sim 1.4$), and in the narrow gap between the two halves of the pixel barrel detector at $\eta \sim 0$. In 2012 data, where tighter tolerances are used in the reconstruction, the efficiency is a a bit lower and the agreement slightly worse, as described in the text.

resources, the efficienies are slightly lower and the agreement is a bit worse: the efficiency measured in data is about 0.4% lower than the expectation from the simulation up to $|\eta| \sim 1.9$, and 1–2% lower in the very forward region. The latter loss of effeciency is mostly due to detector issues localized in space and time which are not yet included in the simulation. The faster decrease of the efficiency in data with respect to simulations for increasing number of primary vertices is under investigation, and is believed to be related to inefficiencies in the pixel detectors at high hit multiplicites and to transient sensor inefficiencies in the strip detectors from highly ionizing particles[5], also proportional to the flux of particles; neither of these effects is currently included in the detector simulation.

2.3.4. Momentum and impact parameter resolution

Two other important indicators of the track reconstruction performance are the resolution on the measured momentum and impact parameter of the tracks. In the context of Higgs searches, the former is especially important for the H → ZZ → 4ℓ search channel at low mass, while the latter allows the suppression of non-prompt backgrounds and an efficient heavy flavour tagging in essentially all Higgs boson search channels.

The momentum scale and resolution have been studied using muons from Z and J/Ψ decays: in the muon reconstruction, the information from the muon system is not used to determine the muon momentum in the p_T range below 200 GeV/c, which is the relevant one for all the SM Higgs boson searches. Two approaches have been used in the measurement, one relying on a parametric description of the Z lineshape (MuScleFit) and one based on shifting and smearing templates from simulation (SIDRA) [30]. The two results are compatible, yielding a momentum resolution $\sigma(p_T)/p_T$ of about 1.4% in the barrel and 2–5% in the endcaps up to $|\eta| \sim 2.1$ (Figure 2.7).

The track impact parameter resolution has been measured using early 2010 data, and found to be in excellent agreement with the expectations from simulations (Figure 2.8) [31]. For transverse momenta above a few GeV/c, when the contributions from multiple scattering become less relevant, the resolution on the transverse impact parameter is independent of η and approaches 20 μm, a figure to be compared with the proper decay length of about 300 μm and 500 μm for D and B hadrons respectively.

---

[5] Very low momentum particles, *e.g.* secondaries from nuclear interactions, can yield large energy losses in the silicon. In these cases, the time needed to restore the nominal bias voltage to the sensor afterwards is longer than the bunch LHC separation, resulting in a hit inefficiency in the subsequent few LHC bunch crossings.

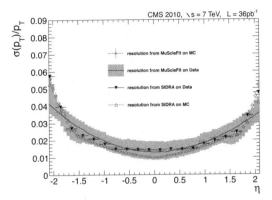

Figure 2.7. Relative transverse momentum resolution $\sigma(p_T)/p_T$ in data and simulation measured by applying the MuScleFit and SIDRA methods to muons produced in the decays of Z bosons.

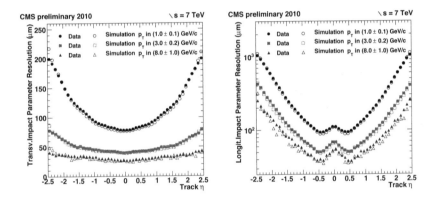

Figure 2.8. Measured resolution of the track impact parameter in the transverse (left) and longitudinal (right) plane as a function of the track η, in three different track p_T ranges.

The longitudinal impact parameter grows exponentially with $|\eta|$ for all momenta, as expected geometrically: the dominant contribution to the resolution in the forward region is from the determination of the polar angle θ, which is measured with an approximately constant absolute uncertainty driven by the hit position resolution in the radial coordinate. At small η, instead, the resolution improves slightly with increasing track angle: particles orthogonal to the sensors yield single-pixel hits for which position resolution is fixed by the granularity of the subdetector, while particles at larger angles yield extended clusters for which a more precise position estimate can be achieved from the measured charge distribution in the pixels of the cluster.

2.3.5. Primary vertex reconstruction

Primary vertex reconstruction is a two-step procedure: tracks are first clustered into sets that appear to come from the same interaction, and then the constituents of each cluster are fitted in order to obtain a precise measurement of the vertex position.

The main challenge for the clustering step in the presence of high pile-up is to avoid merging tracks from separate collisions into a single cluster, while at the same time preserving a good efficiency of assigning tracks to the proper clusters, and a low rate of fake clusters not corresponding to a pp interaction. The algorithm used in CMS to perform clustering is the deterministic annealing [32, 33]: starting from a very loose clustering in which the same track can contribute to multiple clusters, the association between tracks and clusters is iteratively tightened to produce narrower clusters and more definite track-cluster assignments. With this algorithm it is possible to resolve vertices with separations down to about 1 mm, appropriate for a multiplicity of interactions per bunch crossing up to 20, as the longitudinal RMS spread of the luminous region is about 6 cm: as shown in Figure 2.9, the deterministic annealing clustering preserves the linearity between the number of pile-up interactions and reconstructed vertices, while less performing algorithms such as the one used in 2010 data taking result in a loss of efficiency at high multiplicities because of their inability to resolve vertices at too small separations.

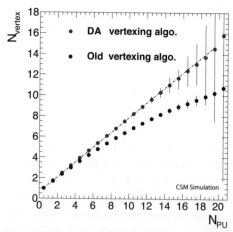

Figure 2.9. Number of reconstructed and vertices as function of number of simulated pile-up interactions, for the deterministic annealing clustering used in 2011 (red) and the old algorithm used in 2010 (black).

Vertex fitting is performed with the adaptive vertex fitter algorithm [34], which addresses the issue of secondaries and fake tracks in the cluster by iteratively downweighting the tracks which are not compatible with the common vertex being fitted. The resolution on the vertex position depends on the number of tracks in the vertex, and is in the 20–50 μm range for the typical multiplicities of the events considered in these searches.

2.3.6. Heavy flavour tagging

The identification of jets from the hadronization of heavy flavour quarks is performed using information from reconstructed tracks, and it relies on the other subdetectors only for the determination of the jet direction. The b-tagging algorithms are based on the presence of tracks with large impact parameter close to a jet and possible secondary vertices reconstructed from those tracks. The two algorithms that were used in the context of the SM Higgs boson searches are briefly described in this section; an extensive description of all algorithms and of the measurement of their performance from data is provided in references [35, 36].

The simplest algorithm, used throughout most of the Higgs boson searches, is the track counting high efficiency tagger (TCHE): for all tracks associated to the jet, the significance of their signed 3D impact parameter S_{IP} is computed with respect to the closest primary vertex and the jet direction[6]. The discriminator variable associated to the jet is defined as the second-largest S_{IP} value. The algorithm requires only the presence of two tracks, and is therefore usable also for very low momentum jets where the geometrical association between jets and tracks becomes inefficient as particles are less collimated; this feature, combined with the possibility of choosing an operating point with very high efficiency for b-jets, makes this algorithm optimal in the context of the $H \rightarrow WW$ analysis for vetoing background events from $t\bar{t}$ and tW production.

The most complex algorithm, used in the $H \rightarrow b\bar{b}$ analysis, is the combined secondary vertex tagger (CSV). This algorithm relies on a multivariate combination of information from all the track impact parameters and possible secondary vertices in the jet. The algorithm is explicitly tuned to provide also a good rejection of jets from c quarks.

Several techniques have been deployed to measure the b-tagging efficiency from data, using inclusive QCD jet events and $t\bar{t}$ events; the separate combinations of the results from the two event samples yield compatible results, with relative accuracies of 4% or better. The mistag rate, *i.e.* the probability for a jet not originating from a b quark to be tagged as

---

[6] The sign of the impact parameter is determined by the sign of the scalar product between the vector connecting the primary vertex to the point of closest approach of the track and the jet direction.

b-jet, has also been measured from the data on QCD jet events, with a relative uncertainty of about 8%. In general, data and simulations are found to be in agreement within about 10% both for efficiencies (Figure 2.10) and mistag rates (Figure 2.11).

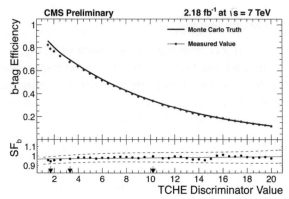

Figure 2.10. Tagging efficiencies measured in $t\bar{t}$ events in data compared to the expectations from simulations, as function of the b-tagging discriminator, for the track counting high efficiency (TCHE) algorithm. The ratio of efficiencies in data and simulation SF_b is displayed in the lower panel, with a smooth parametrization (solid red line) and the overall uncertainty bands (dashed blue lines). The three arrows indicate the loose, medium and tight working points, corresponding to approximately to mistag rates of 10%, 1% and 0.1%.

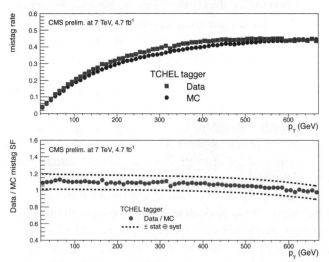

Figure 2.11. Mistag rate for the track counting high efficiency tagger (TCHE) at the loose operating point (L), for data and simulations (top panel) and their ratio (bottom panel). The solid curve in the lower panel is the results of polynomial fits to the data points, while the dashed curve represent the overall statistical and systematic uncertainty on the measurement.

2.4. Muon reconstruction and identification

Muons are a key tool in the CMS physics program, as they can be reconstructed with high efficiency and purity even at the earliest trigger levels, being the only charged particles not stopped by the interactions with the detector material before the muon system.

The design of the CMS detector was indeed done considering the search channel H $\to$ ZZ $\to$ 4 μ as one of the main goals.

The muon detector, trigger, reconstruction and identification were commissioned with cosmic rays in 2008–2009 [37], and with collision data in 2010 [38, 30].

2.4.1. Muon detector layout

The muon tracking system is made of drift tubes ($|\eta| < 1.3$) and cathode strip chambers ($0.9 < |\eta| < 2.4$), inside the steel return joke of the magnet; both devices are self-triggering. For redundancy resistive plate chambers are also installed in most of the detector ($|\eta| < 2.1$) to provide an additional trigger system. A schematic cross-section of the CMS muon system is shown in Figure 2.12.

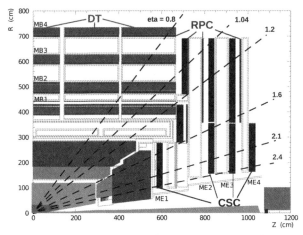

Figure 2.12. Layout of one quadrant of CMS, showing drift tubes (TD), cathodic strip chambers (CSC) and resistive plate chambers (RPC).

Drift tubes (DT).   The barrel drift tubes are organized in four stations composed of two or three stacked superlayers, each consisting of four layers of drift tubes staggered to solve the left/right ambiguities; a honeycomb spacer is inserted between the two superlayers measuring the ϕ coordinate.

The muon position in each DT is reconstructed by measuring the drift time of the ionization electrons, and converting it into a distance from the wire. In all stations, the tubes in the innermost and outermost superlayers have wires parallel to the z axis, to provide accurate measurements of the ϕ coordinate. The intermediate superlayer, present in all but the outermost station, has wires along the ϕ coordinate and provides a measurement of the z coordinate.

Each station can measure the muon position with 100 μm resolution in $r\phi$ and 150 μm in z, and its direction with ~ 1 mrad accuracy.

Cathode strip chambers (CSC).   Drift tubes cannot be used in the endcaps because of the strongly varying magnetic field and the much higher flux of hadron punch-through and radiation (mostly neutrons). Instead, the more robust cathode strip chambers have been chosen.

In these chambers, closely spaced anode wires are stretched between two cathodes, one segmented in strips perpendicular to the wires, and the other without segmentation. A particle traverses the cathodes and produces an ionization trail through the chamber; the electrons are accelerated to the wire, where an avalanche occurs, inducing a charge on the cathode. The coordinate orthogonal to the wire is measured by fitting the charge distribution on the cathode strips, while the longitudinal one is given by the wire that has been hit.

Each station is composed by six layers of cathode strips and wires: the hit point resolution is 75–150 μm in the $r\phi$ coordinate, and 200 μm in the r coordinate.

Energy Loss.   Because of the large amount of material before the muon system, the energy loss by ionization is significant, and very soft muons cannot be reconstructed from the muon system.

On average, muons in the barrel lose about 3 GeV of transverse momentum in the path between the interaction point and the first muon station, and another 3 GeV to reach the outermost station. In the endcaps, the amount of traversed material does not scale with $\sin \theta$, so the p_T loss is dependent on η; roughly, it varies between the p_T loss in barrel and half that value.

2.4.2. Standalone muon reconstruction

The first step for muon identification at CMS is the reconstruction using the muon system alone, to produce track segments or full tracks which can be matched to the tracks reconstructed in the inner detector. The algorithm used is hierarchical: segments are first reconstructed in the individual stations, and then combined to produce tracks.

DT segments.   As the magnetic field is almost completely confined in the steel return joke of the magnet, within each DT station the muon tracks are well approximated by straight lines.

Segments are first reconstructed independently in the transverse and longitudinal planes (r–ϕ and r–z respectively), through pattern matching and linear fitting of the hits, solving also the left/right ambiguities; such 2d segments provide a measurement of one coordinate and one angle.

Afterwards, the two views are combined to produce full "4d" segments, which carry information about two position coordinates and two angles.

CSC.   In each CSC layer, hits are reconstructed by combining the information from anode wires and cathode strips to obtain a 2d point in the transverse plane. Segments are then formed from hits in all the six layers of each station.

Tracks.   In order to reconstruct full muon tracks, the segments are extrapolated from one DT or CSC to the next taking into account the magnetic field and the energy loss and multiple scattering in the steel; if a compatible segment is found in the other station the two measurements are combined and the procedure is repeated for the other stations. Hits reconstructed in the RPC system are also included in the tracks at this stage. The beam spot position in the transverse plane is also used in the standalone muon reconstruction: combined with the measurement of the track angle at the entrance of the muon system, this information allows a muon p_T measurement exploiting the large bending power of the magnetic field inside the solenoid, and not just in the return joke.

2.4.3. Muon identification

Two approaches are used to combine the information from the inner tracker and the muon system in order to reconstruct muons:

- *Global Muon reconstruction*: tracks from the standalone reconstruction are matched to tracks from the inner tracker, and a single Kalman Filter fit is performed to the hits in both subsystems.
- *Tracker Muon reconstruction*: tracks from the inner tracker are extrapolated to the muon system, searching for matched segments in the DT or CSC stations. This approach allows to recover efficiency at low momenta, below about 5 GeV/c, by identifying also muons with hits in a single station, while the standalone and global reconstruction require at least two stations, or one station plus an RPC hit. It also improves the reconstruction of dimuon pairs that have a small separation at the muon system, for which the ambiguity resolution in the standalone introduces inefficiencies.

The large majority of muons of sufficient momentum is reconstructed by both algorithms.

At subsequent analysis stages, additional quality requirements are imposed to the muons to achieve the desired balance between identification efficiency and purity. In this thesis, results for two muon selections will be presented: the loose selection used in the particle flow reconstruction and in the H $\rightarrow$ ZZ $\rightarrow$ 4ℓ analysis, and the tighter selections used in the H $\rightarrow$ WW $\rightarrow$ 2ℓ2 ν search.

PF selection. The identification of muons, including the ones produced within hadronic jets, is very important in the particle flow reconstruction to avoid biases in the jet momenta and E_T^{miss} reconstruction. In this context, even the muons from $\pi^\pm/K^\pm$ in the tracker volume should be identified as muons, as their calorimetric footprint will be that of a minimum ionizing particle. The PF selection applies different requirements depending on whether the muon candidate is isolated or not, and on the associated energy deposition in the calorimeters: (i) isolated global muons are selected first, requiring that the sum of the calorimetric E_T and track p_T in a cone of $\Delta R < 0.3$ centered on the muon candidate does not exceed 10% of the muon p_T; (ii) irrespectively of isolation, muons are accepted at the beginning of the particle flow algorithm if they satisfy tight requirements on the number of hits in the muon system, on the compatibility between the track and the segments, and on the compatibility between the candidate p_T and the footprint in the calorimeters; (iii) at the end of the particle flow reconstruction, muon candidates whose associated calorimetric energy deposition is significantly smaller than the track momentum are accepted if they satisfy looser identification requirements. The muon identification in particle flow is described more in detail in reference [39].

The PF selection is optimal also for the H $\rightarrow$ ZZ $\rightarrow$ 4ℓ analysis, allowing to identify muons with transverse momenta as low as 5 GeV/c, in order to preserve kinematic acceptance also for light Higgs boson masses. With four leptons in the final state, such loose inclusive selection is sufficient to achieve a good purity. Further rejection of non-prompt muons is achieved by means of the isolation and impact parameter criteria.

HWW selection. In the H $\rightarrow$ WW $\rightarrow$ 2ℓ2 ν analysis, a considerably tighter muon selection is applied, justified by the larger branching fraction and the larger reducible background from W + jets: an important difference in reducible backgrounds in the H $\rightarrow$ ZZ and H $\rightarrow$ WW searches is that W + b production is suppressed by non-diagonal terms in V_{CKM} while Z + b$\bar{\text{b}}$ is not. Therefore, reconstructed muons in Z + 2 jets events are more likely to be real muons from heavy flavour decays, while in W + 1 jet a larger fraction of hadronic punch-through or $\pi^\pm/K^\pm$ decays is expected, which can be suppressed applying tighter identification requirements.

The starting point of the selection is the Tight Muon definition [30], which requires the muon to be reconstructed both as a global muon and as a tracker muon, and imposes further quality requirements: the normalized χ^2 of the global track fit less than 10; and at least one hit in the muon system used in the global fit[7]; at least two segments associated to the muon by the tracker muon reconstruction; at least 11 hits in the inner tracker, including at least one in the pixel detector. In the 2012 run, the requirement of at least 11 inner tracker hits have been changed to the requirement of having hits on at least 6 layers in the inner tracker, which was found to be more optimal with the retuning of the tracking algorithms introduced in 2012.

In order to reject muons with poorly reconstructed momenta, which would also result in a mismeasurement of the missing transverse energy, an additional requirement is imposed: the relative uncertainty on the p_T measurement from the Kalman Filter fit must be less than 10%.

A further suppression of muons from $\pi^\pm/K^\pm$ decays in the inner detector is achieved by searching for a "kink" in the reconstructed track: at each layer of the tracker, two separate fits of the track are performed using only the measurements inside and outside the chosen layer, and the muon is rejected if the compatibility χ^2 between the track parameters from the fits is large, favouring the hypothesis that the inner and outer parts belong to the hadron and muon of a $\pi^\pm/K^\pm \to \mu^\pm \nu$ decay.

To improve the efficiency of the selection in the cases where the two muons are at small separation, *i.e.* where the H $\to$ WW is better separated from the irreducible WW background, muons not reconstructed as global muons are also accepted, if they satisfy tighter requirements on the track-to-segment association in the tracker muon reconstruction, and one of the matched segments is in the outermost muon station. This exception to the selection has no impact on the events with a single muon, or two well separated muons.

Selections very similar to the Tight Muon or the HWW one are used also in the other Higgs search channels.

In the context of SM Higgs boson searches, the analyses that rely on the particle flow reconstruction also for leptons use the PF selection as a starting point, on top of which the requirements of the Tight Muon definition are applied to select signal muons.

---

[7] Just like for track reconstruction in the inner detector, hits can be rejected at the fitting stage if they turn out to be incompatible with the fitted track parameters.

Muon identification performance.   The efficiency of muon reconstruction and identification has been measured using the tag-and-probe method on dimuons from J/Ψ and Z decays. Results from the 2011 and 2012 LHC runs for the three selections described above show a good agreement between the performance measured in data and the expectations from simulations in the barrel region, $|\eta| < 1.2$, while in the endcaps the efficiency measured in data is a few percent lower than the expectations (Figure 2.13, 2.14). A better agreement is observed in 2012 data (Figure 2.15).

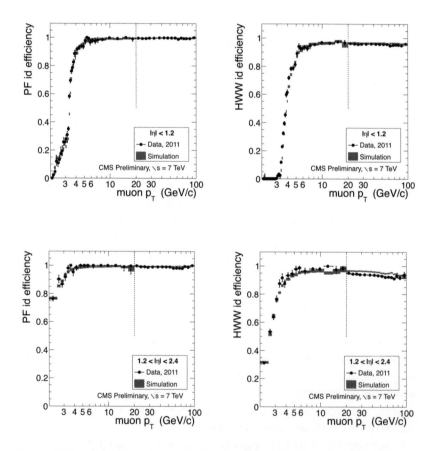

Figure 2.13. Tag-and-probe results for the muon efficiency in 2011 data compared to simulation. Given that a tracker track exists, the plots show the efficiency as a function of muon p_T for the PF (left) and HWW (right) selections in the barrel (top) and in the endcaps (bottom). The measurement is made using J/Ψ → $\mu^+ \mu^-$ events for $p_T < 20\,\text{GeV}/c$, and Z → $\mu^+ \mu^-$ events for $p_T > 20\,\text{GeV}/c$.

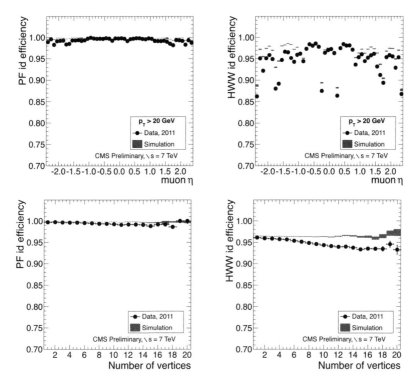

Figure 2.14. Tag-and-probe results for muon efficiency in 2011 data compared to simulation. Given that a tracker track exists, the plots show for the PF (left) and HWW (right) selections as a function of muon η and of the number of reconstructed primary vertices. The measurement is made using muons from $Z \rightarrow \mu^+\mu^-$ decays with $p_T > 20\,\text{GeV}/c$ and $|\eta| < 2.4$.

Part of the inefficiency observed in the endcaps in 2011 is due to chambers that were inactive only in some periods of the data taking, and are therefore not taken into account in the simulation. The remaining is due to an issue in the CSC readout that appeared in the late part of 2011 data taking; as the issue is correlated with high L1 trigger rates, and thus with the instantaneous luminosity, it is also visible in the distribution of the efficiency as function of the number of reconstructed primary vertices (Figure 2.14, bottom row), even if muon identification performance is not expected to be sensitive to pile-up. Both issues have been afterwards resolved, and do not affect 2012 data (Figure 2.15).

2.4.4. Muon isolation

The requirement that a muon is isolated, *i.e.* that the energy flow in its vicinity is below a certain threshold, is very useful in discriminating muons

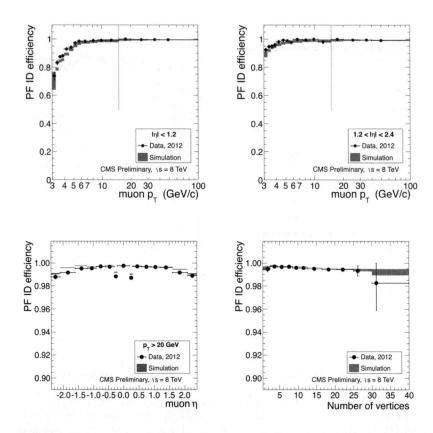

Figure 2.15. Tag-and-probe results for muon efficiency in 2012 data compared to simulation. Given that a tracker track exists, the plots show the efficiency PF selections as function of p_T in the barrel (top left) and in the endcaps (top right), of η (bottom left), and of the number of reconstructed primary vertices (bottom right). The measurement is made using $J/\Psi \rightarrow \mu^+ \mu^-$ events for $p_T < 15\,\text{GeV}/c$, and $Z \rightarrow \mu^+ \mu^-$ events for $p_T > 15\,\text{GeV}/c$.

from decays of W, Z and Higgs bosons from muons produced as a result of QCD processes. Criteria of this kind are used in all SM Higgs boson searches featuring muons.

Energy flow definitions.   Two approaches have been used in the CMS analyses to define the energy flow for isolation purposes: (i) using detector-level information, *i.e.* reconstructed tracks in the inner detector and calorimetric energy deposits in ECAL and HCAL; (ii) using the particles candidates reconstructed from the particle flow algorithm, in which the different subdetectors are combined already. The first approach can benefit from the higher granularity of the information available; the second provides a more transparent treatment of the removal of the muon footprint in the calorimeters, and is less dependent on the performance of the hadronic calorimeter. Eventually, the performances achieved using the particle flow is only very slightly better, so that the choice in each analysis is usually driven by other factors.

In most CMS analyses, the energy flow has been computed by adding up linearly the transverse momenta or transverse energies measured in a cone in (η, ϕ) space, *i.e.* $\sqrt{\Delta\eta^2 + \Delta\phi^2} \leq R$, centered on the muon. To further improve the discrimination between signal and background, in the H $\rightarrow$ WW $\rightarrow$ $2\ell 2\nu$ analysis a multivariate isolation discriminator has been computed combining the information about the energy flow in multiple concentric rings in (η, ϕ), separately for charged hadrons, neutral hadrons and photons, thus exploiting the different distribution of the energy deposits within the isolation cone. This approach allowed a substantial increase in background rejection for the tight isolation criteria used in this analysis.

Pile-up corrections.   In the 2011 and 2012 LHC runs, data was collected with a mean of 5–30 interactions per bunch crossing, potentially giving a large contribution to the energy flow used for isolation and a corresponding loss in signal efficiency. When measuring the energy flow from tracks or reconstructed charged hadrons, which have an associated track, the contributions from pile-up interactions is easily dealt with by considering only tracks compatible with the same primary vertex of the muon; up to minute details on how the compatibility is defined, this approach was used throughout all the SM Higgs boson searches at CMS. Different choices have instead been adopted for the more complex issue of dealing with the contributions from pile-up to the calorimetric deposits, or to the neutral particles reconstructed from the particle flow:

- *No correction*: when a loss in signal efficiency is more acceptable than a loss in background rejection, *e.g.* as in the case of the H $\rightarrow$ WW analysis, the energy flow from neutral particles is not corrected for pile-up effects. A reduced pile-up dependency can be achieved by using particle flow reconstruction, so that the calorimetric footprint of charged hadrons from pile-up is already subtracted, and by using higher energy thresholds on the neutral particles.

- *Energy density corrections*: a per-event median transverse energy density ρ is computed using the FASTJET algorithm [40, 41], and a term proportional to ρ is subtracted from the energy in the isolation cone.
  In principle, it would be natural to use the same energy flow definition used to compute ρ and the isolation sum, but for practical considerations in the Higgs analyses the ρ was always computed from

all particle flow candidates in the $|\eta|$ acceptance for leptons; this is equivalent to the extent in which all energy flows are linearly proportional[8], and in any case the dominant uncertainty on the correction is from the statistical fluctuations of the energy deposit in the cone with respect to the median energy in the event.

The coefficient multiplying ρ used to perform the subtraction can be either the geometric area of the cone πR^2 or an "effective area", *i.e.* an ad-hoc coefficient chosen precisely to achieve an isolation efficiency independent from pile-up.

- *In-cone pile-up corrections*: a different technique to statistically subtract the deposits from neutral particles in the isolation cone was used in the H $\rightarrow$ $\tau\tau$ and H $\rightarrow$ ZZ $\rightarrow$ $2\ell2\tau$. In this technique, the energy flow from charged particles associated to pile-up primary vertices in the isolation cone is used to estimate the corresponding flow from neutral particles. As the ratio of charged to neutral energy in QCD events is approximately two, half of the energy charged energy from pile-up is subtracted from the total neutral energy in the cone, truncating negative results to zero.

  The fundamental difference between this approach and energy density one is in the fact that only the charged energy from pile-up in the cone is used: this leads to a larger statistical uncertainty on the correction, as the pile-up energy flow is sampled in a narrower region, but also to a better correction in the cases where contamination comes from collimated jet-like activity.

2.4.5. Muon trigger

Level 1 muon trigger.   At level 1 the only available information is from the muon chambers and the calorimeters, with reduced granularity. This allows for a bottom-up track reconstruction in the muon system starting from local segments in each station and proceeding to full tracks, processing in parallel different stations or detector slices in the trigger electronics. Apart from the technical implementation and the level of approximations involved, the algorithm is conceptually similar to the one used also offline for the standalone muon reconstruction.

Muons candidates are required to traverse at least two stations, and must have p_T above some threshold dependent on the specific trigger path. Different quality criteria are defined, depending on the number of stations used and if the candidate is confirmed by both the DT or CSC

---

[8] This is true only approximately, as the energy flow in the calorimeters is sensitive also to out-of-time pile-up, while the energy flow in the tracker is not.

and the RPC; in general, tighter requirements are applied to the single muon trigger paths with respect to double or triple muon paths. Precise determination of the p_T at L1 is challenging in the very forward region, $2.1 < |\eta| < 2.4$; because of this, in the last period of 2011 data taking and in 2012 the single muon triggers were limited to $|\eta| < 2.1$, as in the design specifications of CMS.

The L1 electronics also provides to the global L1 trigger a map of the calorimeter regions where the energy deposit is small, so that isolation criteria could be applied to the muons, but this has not been used so far.

High level trigger.   The first step of the muon high level trigger is a more accurate standalone muon reconstruction, thereafter referred as L2, starting from the seeds given by the L1 trigger. This already reduces the input rate significantly, allowing then the reconstruction in the inner tracker to be performed in a cone around the L2 muon, and finally the global muon reconstruction (L3) to further improve the p_T resolution and reject soft muons whose p_T was overestimated by the less accurate L2 reconstruction.

The reconstruction in the inner detector is first attempted outside-in, to benefit from the lower occupancy of the outer tracker. If for a given L2 muon this approach fails, *e.g.* due to geometrical or detector inefficiencies, a second attempt is made inside-out starting from the pixel detector.

In the last part of the 2011 run, a special version of tracker muon reconstruction was implemented in the dimuon trigger paths, to improve the efficiency especially in the endcaps. The strategy for dimuon triggers using also tracker muon is: (i) events are selected starting from an asymmetric L1 trigger with no p_T and quality requirements on one of the two candidates; (ii) the normal HLT reconstruction is performed, and the event is rejected unless at least one high p_T L3 muon candidate is found; (iii) track reconstruction is performed in the full detector, starting from pixels, and the resulting set of tracks is merged with the ones already reconstructed at the L3 step; (iv) tracker muon reconstruction is performed using merged track collection and the muons segments from L2 reconstruction; (v) the event is accepted if two L3-or-tracker muons are found, satisfying the p_T thresholds of the trigger path.

Isolation criteria are also applied in some muon HLT triggers: calorimetric isolation requirements are applied after L2 reconstruction, and isolation using tracks reconstructed in the pixel detector is computed after L3 reconstruction. In the 2012 LHC run, pile-up corrections to the isolation have also been implemented in the High Level Trigger.

Performance.   In the 2011 LHC run, different p_T thresholds have been used to cope with the increasing instantaneous luminosity. The turn-on curves for each time period have been measured on data, using the tag-and-probe method on Z → $\mu^+ \mu^-$ events recorded using a single muon trigger. Trigger thresholds have been stable throughout the first half of 2012, as the increase in instantaneous luminosity has been more modest.

Two example turn-on curves for the logical OR of the single muon triggers with and without isolation requirements are shown in Figure 2.16. As the isolation requirements applied at trigger level do not match precisely with those applied in the offline analysis, the efficiency at the plateau for the trigger without isolation requirements is higher even for muons satisfying the offline isolation requirements.

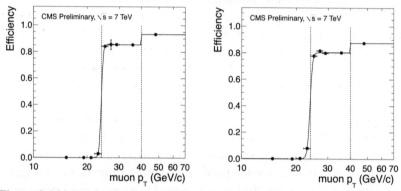

Figure 2.16. Tag-and-probe results for the single muon trigger efficiency as a function of muon p_T, in a period of 2011 when the p_T thresholds at HLT were 24 GeV/c and 40 GeV/c for the triggers with and without isolation requirements. The efficiencies are measured for muons satisfying the HWW selection, for the two rapidity ranges $|\eta| < 1.2$ (left) and $1.2 < |\eta| < 2.1$ (right). The red line shows the parametrization of the efficiency used in the analysis.

The efficiency of the double muon trigger have also been measured with the tag-and-probe method, relying on the fact that the two muons are reconstructed independently, so that the per-event efficiency is simply the product of two per-muon efficiencies. As the single muon trigger requirements are always tighter than the ones of the double muon trigger, it is possible to measure the per-muon efficiency of the double muon trigger by selecting events in which the tag muon satisfies the single muon trigger: the per-muon efficiency, as function of the kinematics of the probe muon, is equal to efficiency for those events to satisfy the double muon trigger.

The per-muon efficiency for the low p_T muon of the asymmetric double muon trigger with p_T thresholds of 17 and 8 GeV/c is shown in Figure 2.17, both when using only the L2+L3 reconstruction and when allowing also for tracker muons at HLT. The gain in efficiency from the latter reconstruction is about 2% in the barrel, 6% in the endcaps, and 10% for $|\eta| > 2.1$.

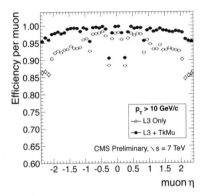

Figure 2.17. Per-muon trigger efficiency for the low p_T muon of the asymmetric double muon trigger with p_T thresholds of 17 and 8 GeV/c, as a function of muon pseudorapidity. The black filled symbols are for the trigger using also the tracker muon reconstruction for the low p_T muons, while empty blue symbols refer to the trigger using only the standard L2+L3 outside-in reconstruction. The efficiency is measured for muons with $p_T > 10$ GeV/c satisfying the H $\to$ WW selection.

2.5. Electron and photon reconstruction and identification

An efficient and precise reconstruction of electrons and photons is crucial in many of the searches for a SM Higgs boson at CMS: the searches in the H $\to$ $\gamma\gamma$ channel, which provide the best sensitivity for a light Higgs boson, have been used as a guiding principle to design the CMS electromagnetic calorimeter; electrons from W $\to$ e ν, Z $\to$ ee, and $\tau \to$ e $\nu\overline{\nu}$, are used in all other search channels alongside the corresponding final states with muons to double or quadruple the phase space.

The experimental challenges in the reconstruction of electromagnetic objects are very different from the ones for muons: the efficiency of detecting energy deposits in the ECAL is 100%, thanks to its hermetic design, but contrarily to muons there is a huge background from the energy deposits of hadronic origin and from $\pi^0 \to \gamma\gamma$. In addition, photon conversion and electron hard bremsstrahlung in the detector material (possibly followed by conversions) are likely events, resulting in complex topologies of energy deposits in the calorimeter.

Achieving a good energy resolution is likewise more challenging for electrons and photons, requiring multiple levels of energy corrections, some also time-dependent.

2.5.1. Electromagnetic calorimeter

The ECAL is a homogeneous calorimeter, made of nearly 76000 scintillating crystals of lead tungstenate (PbWO$_4$), with a pseudorapidity coverage $|\eta| < 3.0$, separated into a barrel region (EB) and an endcap region (EE); a lead/silicon-strip preshower detector is also installed for $1.6 < |\eta| < 2.6$ (Figure 2.18).

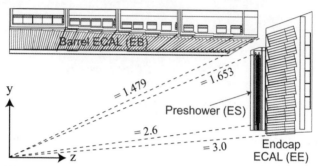

Figure 2.18. Longitudinal view of the electromagnetic calorimeter and preshower detector.

Lead tungstenate is characterized by a radiation length of $X_0 = 0.89$ cm, allowing a very good longitudinal shower containment with a detector depth of about $26 X_0$ in the barrel, and about $(3 + 25) X_0$ in the preshower+endcaps. The small Moliere radius $R_M = 2.2$ cm and the fine transverse segmentation of 2.2 and 2.9 cm in the EB and EE is important for the electron and photon identification using shower shapes and for the performance of particle flow reconstruction of taus and collimated jets.

Crystals are read out using avalanche photodioes (APD) in the barrel, and vacuum phototriodes (VPT) in the endcaps.

2.5.2. Electron and photon reconstruction

Electron and photon reconstruction in the ECAL is performed by grouping channels into clusters corresponding to single showers, which are then merged into more extended superclusters designed to collect the energy of an electron or photon and all its associated radiation from bremsstrahlung photons and conversion tracks. Due the strong magnetic field and the non-negligible amount of detector material upstream to the calorimeters (Figure 2.19), calorimetric energy deposits are narrow along the η direction but can be large in the ϕ direction, especially in the barrel and at low momenta.

Reconstruction of electron tracks, both for primary electrons and for photon conversions, is performed by searching for track seeds compatible with the ECAL superclusters. As electrons can undergo substantial

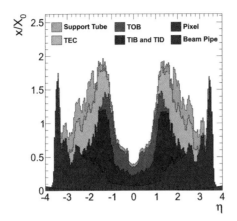

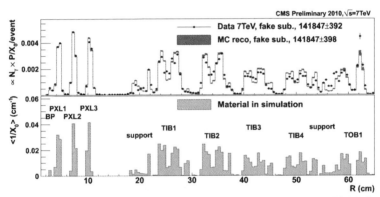

Figure 2.19. Up: material budget in the tracker upstream to the ECAL in units of radiation length as a function of η, extracted from simulation; down: material budged in the tracker as function of the radius R, measured using reconstructed photon conversions in data and simulation (top panel) and extracted directly from simulation (bottom panel) [42].

energy losses and large angle scatterings in the material, large search windows are used in the pattern recognition stage, and the final track fit is performed using the Gaussian sum filter (GSF) algorithm which properly accounts for these effects [43]. This track reconstruction algorithm allows also a measurement of the energy loss due to radiation by comparing the local curvatures of the trajectory at the two endpoints.

The reconstruction algorithm based on ECAL superclusters is complemented by a particle-flow based one where electromagnetic energy deposits are searched for along tangents to the electron candidate tracks drawn in correspondence to the tracker layers, where most of the detector material is located. The combination of the two algorithms significantly improves the efficiency for electrons that have low momentum or are not isolated.

An illustration of these features of the electron reconstruction is shown in Figure 2.20. More detailed information about the electron and photon reconstruction algorithms and results from the commissioning with the first collision data can be found in references [44, 45, 46, 39].

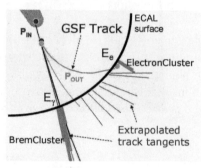

Figure 2.20. Illustration of an electron undergoing a large radiative energy loss in the material, depicted in the transverse plane: the local curvature of the GSF track close to the interaction point provides an estimate of the initial momentum p_{in}, while that close to the ECAL surface allows to estimate the final one p_{out}. The tangents to the trajectory are used to collect the bremsstrahlung clusters in the particle flow.

2.5.3. Electromagnetic energy calibration

The energy measurement from the ECAL is used to determine the energies of reconstructed photons, and dominates the combination of tracker and ECAL used for electrons for p_T above about 20 GeV/c, especially for electrons that undergo large energy losses in the tracker material.

The energy calibration at the level of individual detector channels is performed using the azimuthal symmetry of the energy flow, with corrections for the displacement of the beam spot in the transverse plane, and photons from π^0 decays. Diphotons from π^0 decays also provide a first absolute energy scale calibration. This first calibration is also validated using electrons from W $\rightarrow$ e ν and Z $\rightarrow$ ee. Especially in the endcaps, the irradiation during the LHC running causes transient losses of transparency, recovered during the periods with no collisions; in order to continuously monitor the evolution of the transparency, a monitoring system is set up using light injected from lasers and LEDs.

A second layer of energy calibrations is applied at the level of the reconstructed supercluster. Two complementary approaches have been used: (i) a traditional one, based on factorized corrections for shower containment and energy losses in the detector material as function of the supercluster E_T and η, and its size along ϕ, the latter variable being strongly correlated with the amount of radiative energy loss; (ii) a novel approach based on a multivariate regression trained on simulated photons, using as input variable the global η and ϕ of the supercluster, a collection of shower shape variables and a set of local cluster coordinates. The latter approach, developed specifically for the 2011 H $\rightarrow$ $\gamma\gamma$ analysis, was found to improve substantially the energy resolution, especially in the EB where the detector geometry is more accurately modelled in the simulation.

Finally, the absolute energy scale is calibrated from a fit to the Z $\rightarrow$ ee lineshape, separately in four $|\eta|$ regions; this provides a measurement in a kinematic regime similar to that of the Higgs boson signals, without the large extrapolations that would be needed to use the π^0 decays. This calibration procedure is also used to determine from data the energy resolution (Figure 2.21).

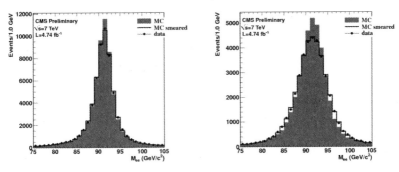

Figure 2.21. Invariant mass distribution of electron pairs from Z $\rightarrow$ e$^+$e$^-$ decays, where both electrons are in the central barrel (left), and where one electron is in the central barrel and and the other is in the endcap (right). The red histogram is Monte Carlo simulation, and the black empty histogram is the same simulation but after a smearing of the energies to match the data.

2.5.4. Electron and photon identification

Since practically all particles produced in the pp collisions that reach the ECAL will result in an energy deposit, identification requirements are of paramount importance to separate the signal electrons and photons from the reducible backgrounds. Suitable discriminating variables are the ratio between the energy detected in ECAL and in HCAL, and the transverse size of the shower in the η coordinates, which is insensitive to radiative losses.

In the case of electrons, additional discriminating variables are the ratio between the energy measured in the tracker and in the ECAL, and the distance between the barycenter of the supercluster and the impact point of the GSF track extrapolated to the ECAL surface. A further improvement in discrimination for electrons is achieved by using two indicators of radiative energy losses, expressed as the ratio between estimates of the electron momentum at the production point and at the ECAL: for both, the initial momentum is estimated using the local curvature at innermost point of the GSF track; the final momentum is estimated either using the local curvature at the other endpoint of the GSF track, or using the most energetic cluster within the supercluster.

Identification of electrons from photon conversions in the tracker is also important: when selecting prompt electrons, it is needed to reject $\pi^0 \to \gamma\gamma$ followed by $\gamma \to e^+ e^-$, for which the previously mentioned variables provide scarce rejection power; conversely, when selecting prompt photons, it is needed to preserve the efficiency for converted photons while rejecting other electrons. Two criteria are used: (i) tracks from conversion electrons should not have hits in the innermost tracker layers, upstream to the conversion, while tracks from prompt electrons are supposed to have hits whenever their trajectory crosses the sensitive area of an active tracker module; (ii) conversions in which both electrons are reconstructed in the inner tracker can be identified by the presence of two oppositely charged tracks with collinear momenta at the conversion vertex.

Information about the impact parameter of the electron tracks with respect to the primary vertex can also be used to discriminate against electrons from the decays of heavy flavour hadrons and from photon conversions.

Different approaches are used in the various analyses to select electrons and photons on the basis of these variables, ranging from simple selection criteria applied to each variable individually to the use of multivariate classifiers combining all or some of these variables into a single discriminator.

2.5.5. Isolation

Just like for muons, isolation is a powerful tool to separate signal photons and electrons from the reducible backgrounds. Likewise, the energy flow definitions and techniques to control the contaminations from pile-up are the same as for muons, described in Section 2.4.4.

One aspect that sets electrons and photons apart from the muons is however the removal of the footprint from the isolation sum, which is made more complex by the fact that bremsstrahlung photons and tracks from conversions can contaminate a region with a large extension along the ϕ direction, so that it is no longer sufficient to ignore only the deposits almost collinear with the electron or photon. When isolation sums are computed from energy deposits in the ECAL, or the photons reconstructed from the particle flow algorithm, this issue is dealt with by applying a veto to objects that are close in η to the electron, irrespectively of the distance along ϕ.

In some analyses, a similar approach is used also for the isolation computed from tracks, in order to reject electrons from photon conversions, while in others the tracks from late photon conversions are removed by vetoing tracks with large impact parameter[9]. When computing isolation from the output of the particle flow algorithm, this veto can be restricted only to electron candidates, preserving the discriminating power against background.

2.5.6. Triggers for electrons and photons

Isolated electrons and photons can to some extent be discriminated from the jet-induced backgrounds on the basis of calorimetric information alone, allowing these objects to be selected already at the earliest trigger levels.

L1 e/γ trigger.   In the L1 trigger, ECAL crystals are grouped into trigger towers, fixed arrays of 5×5 crystals in the EB and with a more complex geometry in the EE, in both cases designed to match the coarser granularity of the HCAL. Individual ECAL trigger towers are selected as electron candidates if the shower is contained in a narrower array of 2×5 crystals along (η, ϕ), and if the energy deposit in the corresponding HCAL tower does not exceed a given fraction of the ECAL energy. In order to improve the containment, the energy of the candidate is estimated including also the tower with the highest energy among the four adjacent ones.

As no information from the inner tracker is available at L1, no charge determination is possible, and so electrons and photons are indistinguishable at this stage.

Electron isolation criteria at L1 can in principle be applied by requiring negligible activity in neighbouring towers, but just like for muons this feature has not been used in the 2010 and 2011 LHC runs.

High level trigger.   ECAL reconstruction in the HLT is performed in a similar way as in the offline except that superclusters are required to be matched with L1 e/γ candidates, which also allows to save processing time since only the detector channels in the neighbourhood of L1 candidates have to be analyzed. At lower integrated luminosities, this require-

---

[9] Early photon conversions result in tracks collinear with the photon at the interaction points, which are therefore already removed by vetoing collinear deposits. In general, rejecting tracks with large impact parameter can have a negative impact on the discrimination against jet-induced background, due to the loss of detached tracks from heavy flavour decays, long-lived neutral particles, nuclear interactions and conversions electrons from $\pi^0 \to \gamma\gamma$ decays.

ment has been relaxed for dielectron and diphoton triggers, allowing one of the two objects to be found at a second stage of HLT processing when the reduced event rate makes it possible to process information from the whole detector readout.

Calorimetric photon and electron identification requirements based on the shower shape can then be applied to the reconstructed superclusters; likewise, isolation criteria on the basis of the energy deposits in the ECAL and HCAL can be used to suppress jet-induced backgrounds at this trigger stage.

Electron track reconstruction with the full GSF algorithm is too computationally intensive to be used at HLT. A faster track reconstruction is performed instead, seeded by pairs of hits in the pixel and innermost strip detector layers and using the normal Kalman Filter algorithm; while the KF cannot reconstruct the full electron track in case of large radiative energy losses, it can still provide a measurement of the electron momentum at the interaction point, sufficient for applying simple electron identification requirements. The use of GSF tracking at HLT which would allow also more complex electron identification requirements, is being considered for the 2012 run.

After the event rate has been reduced by the previous selection stages, track reconstruction in the neighbourhood of the electron or photon candidates can be performed, allowing also for isolation criteria based on reconstructed tracks.

Performance.   In general, given the very high degree of correlation between the online and offline reconstruction and identification algorithms, electron and photon triggers are characterized by plateau efficiencies of 99% or higher for candidates satisfying offline analyis requirements. The sharpness of the turn-on, instead, can be affected by the different quality of reconstruction and calibrations used at the HLT with respect to the offline (Figure 2.22). This effect is more visible in the endcaps, where time-dependent transparency losses have a larger impact; usage of the transparency calibrations also in the endcaps is foreseen for the 2012 run.

2.6. Particle flow reconstruction

The CMS particle-flow algorithm aims at an exclusive reconstruction and identification of all stable particles in the event, *i.e.* electrons, muons, photons, charged hadrons and neutral hadrons, by means of an optimized combination of information from all subdetectors. The algorithm is described in detail in reference [47], and information on its commissioning with early data are provided in references [48, 49, 39].

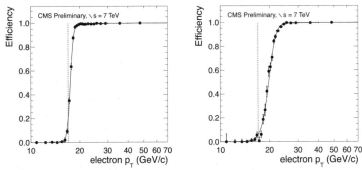

Figure 2.22. Tag-and-probe results for the high momentum leg of the double electron trigger, with a p_T threshold of $17\,\mathrm{GeV}/c$, in the last part of the 2011 run. The efficiencies are measured on from $Z \rightarrow e^+e^-$ decays, for electrons satisfying the selection used in the $H \rightarrow WW$ analysis, in the barrel (left) and endcaps (right). The red line shows the parametrization of the efficiency used in the analysis.

The core of the particle-flow reconstruction technique is the algorithm used to link the signals in the individual subdetectors, *i.e.* tracks or calorimetric clusters, whenever they can possibly originate from the same particle. The association used in the linking stage is purely geometrical: tracks are linked to calorimetric clusters if their trajectory intersects one of the calorimetric cells of the cluster; and likewise clusters in the ECAL preshower, ECAL and HCAL are linked if the cluster position measured in the finer granularity subdetector lies within the envelop of the cluster in the coarser granularity subdetector. In order to account for uncertainties from multiple scattering in the track extrapolation and on the estimated position of the shower maximum in the calorimeters, a geometrical tolerance of the size of one calorimeter cell is included when defining links; this tolerance can also account for gaps and cracks in the calorimeters. By design, the linking algorithm is simple and robust, as it does not rely on the precise knowledge of the position resolution in each subdetector. Specialized algorithms are used for linking tracks to recollecting bremsstrahlung clusters in the case of electrons by considering tangents to electron trajectories at the crossing points with the tracker layers, as described in Section 2.5.2.

Blocks of one or more linked objects are then processed to identify and reconstruct particle candidates. Isolated electrons and muons are selected first, and reconstructed using the dedicated algorithms developed for them, as described earlier in Sections 2.4.3 and 2.5.2; similarly, non-isolated tracks which satisfy tight muon identification criteria are immediately identified as muons. Charged hadrons are identified as tracks in the inner tracker, normally linked to calorimetric deposits if the particle p_T is sufficient for the trajectory to reach the calorimeters. If the momentum measurements from the track and calorimeter are compatible, after accounting for non-linearities and zero suppression effects, the best energy determination is obtained as a combination of the two. If the track momentum significantly exceeds the measured calorimetric energy, the particle is identified as muon if it satisfies very loose muon identification criteria; otherwise, tight track quality requirements are applied to to reject mis-reconstructed tracks. If instead an excess of calorimetric energy deposition is found with respect to the momentum of the associated track, or the associated tracks if the same cluster is linked to multiple ones, *e.g.* in the case of collimated hadronic jets, the residual energy is identified as a photon or a neutral hadron. Additional photons and neutral hadrons are also identified from calorimetric deposits not linked to any track.

2.6.1. Jet and E_T^{miss} missing transverse energy reconstruction

Jet reconstruction is defined by a set of objects, and an algorithm to cluster them; several options are available in the CMS event reconstruction for both these choices but the vast majority of the analyses, including all the SM Higgs boson searches, rely on jets from particles reconstructed using particle flow approach and clustered according to the anti-k_T algorithm with distance parameter $R = 0.5$.

The particle flow approach, exploiting the precision of the inner tracker and the full granularity of ECAL, provides a very significant improvement to the jet energy and direction resolutions compared to calorimetric-only approaches, as charged hadrons and photons account for about 90% of the energy in jets. The use of particle flow reconstruction also allows a high reconstruction efficiency for jet constituents with low momentum, thereby achieving a jet energy response close to unity already before calibrations.

In some analyses, particle flow objects corresponding to isolated and well-identified leptons, and charged hadrons associated to vertices other than the primary one, are removed prior to the jet clustering. In most of the analyses, however, the clustering is done inclusively on all objects. The performances obtained with the two approaches are anyway very similar, as long as jet energy corrections for pile-up are applied consistently in the two cases.

Jet clustering. The anti-k_T algorithm [50] is a sequential recombination algorithm like k_T and Cambridge/Aachen, in which objects are clustered hierarchically starting from the pair with smallest distance

$$d_{ij} = \min\left(p_T^{-2}(i),\ p_T^{-2}(j)\right)\frac{(\eta_i - \eta_j)^2 + (\phi_i - \phi_j)^2}{R^2},\qquad(2.1)$$

and clusters are promoted to final jets when their p_T^{-2} is smaller than any remaining distance d_{ij}. As the other sequential recombination, it has the property of yielding stable results in the presence of collinear or infrared radiation, necessary for well defined theoretical calculations at parton level beyond leading order.

The use of p_T^{-2} in the distance and promotion condition is what sets this algorithm apart from k_T and Cambridge/Aachen, which use respectively p_T^2 and p_T^0 (*i.e.* a constant). The important consequence of this difference is that with anti-k_T soft radiations around hard objects is clustered first, since the distance d_{ij} is downscaled by the square of the p_T of the hard object; and since the distance does not depend on the p_T of the soft object the algorithm produces conical jets of radius equal to the distance parameter, unless there are multiple hard objects separated by less than R. This is advantageous experimentally, as it simplifies the event interpretation and the correction of effects from pile-up.

Jet energy resolution. Dijet and $\gamma +$ jet events in data also been used to determine the jet energy resolution. In general, the energy resolution measured from data is found to be close, but slightly worse, than the predictions from simulation. The resolution improves with the jet p_T, typical values being about 15% at 30 GeV/c and 10% at 100 GeV/c (Figure 2.23).

Energy scale calibration. CMS uses a factorized approach to jet energy calibration, as described in detail in reference [51]. First, an offset correction is applied to account for contamination from pile-up, determined from the per-event median energy density computed with the k_T algorithm and the jet areas [40, 41], modulated as function of η to allow for the non-uniformity of the energy flow and detector response. Then, corrections for jet energy response as function of η and p_T; a first level of corrections are derived on the basis of simulated events, on top of which residual corrections for differences in the response between the real detector and simulation are obtained from dijet, $\gamma +$ jet and Z + jet events in data. The overall uncertainty on the jet energy scale for jets with $p_T > 30$ GeV/c is better than 3% in the barrel, and better than 5% up to $\eta \sim 4.5$.

Pile-up jet tagging. Several Higgs searches rely on hadronic jets for event categorization, in particular when targeting the VBF production

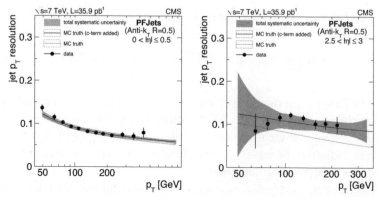

Figure 2.23. Left: jet energy resolution for anti-k_T jets obtained from particle flow objects, measured from dijet events in data, in two pseudo-rapidity regions. The resolution expected in simulations is shown as a dotted red line; the best estimate of the resolution in data obtained adding a constant term to the predictions from simulations is shown as a solid red line, with the associated systematical uncertainty band [51].

mode. With the increase of pile-up in the 2012 running period, an algorithm has been deployed to identify the hadronic jets arising from pile-up activity, also outside the coverage of the tracker where the association of charged hadrons constituents to the individual primary vertices cannot be used. The algorithm relies on the different angular distribution of the jet constituents, more collimated for signal jets than for pile-up jets which arise mostly mostly from combinatorial coincidence of uncorrelated particles; for central jets, tracking information is also used. The performance of the algorithm has been validated in $Z \to \mu^+ \mu^-$ events: after removing jets tagged as pile-up, the jet multiplicity was found to be stable within better than 1% for jet p_T thresholds down to 20 GeV/c.

Missing energy reconstruction. In the context of particle flow reconstruction, the missing transverse energy is defined as the negative vector sum of all reconstructed particles in the event. As for jets, the performance achieved with this approach is vastly superior to the one obtainable from calorimetric energy measurements alone.

Detailed studies of the missing energy performance have been done with the data collected in the 2010 LHC run [52]. A very good agreement between data and predictions from simulation is observed in events with no pile-up, both when the E_T^{miss} is purely instrumental, *e.g.* in γ + jets, and when genuine momentum unbalance from neutrinos is present, *e.g.* $W \to \ell\nu$ events (Figure 2.25).

In the higher pile-up environment of 2011 data taking, a degradation of the E_T^{miss} resolution in events with no genuine missing energy is ob-

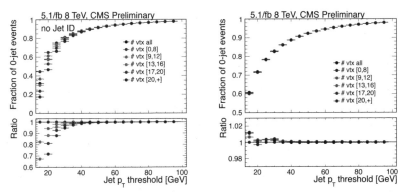

Figure 2.24. Fraction of Z → μ⁺ μ⁻ events with zero hadronic jets as function of the jet veto threshold, for different multiplicities of primary vertices. In the left plot, all jets are considered, while in the right plot the jets tagged as pile-up are removed.

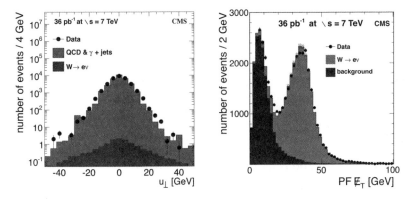

Figure 2.25. Missing transverse energy in γ + jets events (left) and in W → e ν events (right). In the left plot, only events with a single reconstructed primary vertex, and the $\vec{E}_T^{miss}$ is projected in the direction perpendicular to the photon $\vec{p}_T$ [52].

served. As a consequence, several approaches to improve the rejection of low-E_T^{miss} backgrounds have been pursued in the different analyses, mostly based on using information from charged particles associated to the selected primary vertex; the specific choices adopted in the H → WW analysis are described in Section 3.3. The degradation is found to be only partially modelled by simulated events[10], so techniques have been developed to improve the description of simulated backgrounds by means of embedding simulated particles into reconstructed events from data (as

---

[10] This can be due to the limited accuracy of the modelling of minimum bias events.

done *e.g.* for $Z \rightarrow \tau^+\tau^-$) or correcting the simulated missing energy using recoil information measured in $Z \rightarrow \ell^+\ell^-$ events.

2.6.2. Tau reconstruction and identification

The branching fraction for tau lepton decays are approximately one third into lighter leptons and neutrinos ($\tau^- \rightarrow \ell^- \overline{\nu}_\ell \nu_\tau$) and two thirds into hadrons and a ν_τ; the hadronic decays are dominated by modes with one or three charged mesons (mostly $\pi^\pm$) and up to two neutral pions, often mediated by intermediate resonances such as the ρ and a_1 mesons (Table 2.1). While not much can be done for leptonic tau decays, dedicated reconstruction algorithms for hadronic tau decays have been developed in CMS [53].

one charged		one charged $+ n\,\pi^0$		three charged	
$\tau^- \rightarrow e^- \overline{\nu}_e \nu_\tau$	17.9%	$\tau^- \rightarrow \pi^- \pi^0 \nu_\tau$	25.5%	$\tau^- \rightarrow \pi^- \pi^- \pi^+ \nu_\tau$	9.3%
$\tau^- \rightarrow \mu^- \overline{\nu}_\mu \nu_\tau$	17.4%	$\tau^- \rightarrow \pi^- 2\pi^0 \nu_\tau$	9.3%	$\tau^- \rightarrow \pi^- \pi^- \pi^+ \pi^0 \nu_\tau$	4.6%
$\tau^- \rightarrow \pi^- \nu_\tau$	10.9%	others with one π^0	0.5%	others with no π^0	0.5%
$\tau^- \rightarrow K^- \nu_\tau$	0.7%	others with two π^0	0.2%	others with $\geq 1\pi^0$	0.5%
		modes with $\geq 3\pi^0$	1.4%		

Table 2.1. Tau decay branching fractions for various final state topologies [54]. The $\pi^-\pi^0$ decay mode happens dominantly through the $\rho^\pm$ resonance, while the 3π decays are mediated by the $a_1^\pm$ resonance. Fractions don't add up to 100%, as they are result of separate measurements with independent uncertainties of $O(0.1\%)$, and some decay modes are not listed.

Reconstruction and isolation. All tau reconstruction algorithms in use at CMS start from jets reconstructed from the output of the particle-flow algorithm, clustered using the default anti-k_T clustering with $R = 0.5$. A tau decay mode analysis is then performed, reconstructing π^0 mesons from photons candidates, combining them with charged hadrons to reconstruct the full decay mode and tau four-momentum; an isolation criteria is then applied using all other particles in the jet.

In the majority of the analyses, including all SM Higgs boson searches, the hadron plus strip (HPS) reconstruction algorithm is used. In this algorithm, photons candidates are clustered in strips of size $\Delta\eta = 0.05$ and $\Delta\phi = 0.20$, centered first on the most energetic photon, and then iteratively re-centered on the total four-momentum of the cluster. Up to two strips, each with $p_T > 1$ GeV/c are used, although in the largest fraction of the decays only one strip is found.

Four final state topologies are considered: a charged hadron with zero to two strips, and three charged hadrons. The reconstruction is attempted in each topology, and if more than one yields a successful tau candidate the one with largest transverse momentum is selected.

In the case of a single charged hadron, nothing has to be done. About 25% of the hadronic tau decays are reconstructed only in this topology, arising mostly from genuine decays to a single hadron (60%) or from decays involving non-reconstructed π^0 mesons (35%).

When a single strip is found, a π^0 candidate is reconstructed from the total momentum of the photons in the strip and assuming the π^0 mass value. An attempt is then made to pair the π^0 with a charged hadron to reconstruct a $\tau \to \rho\, \nu_\tau \to \pi\,\pi^0\,\nu_\tau$ decay chain, imposing a loose mass constraint $0.3 < m_\rho < 1.3\,\mathrm{GeV}/c^2$; the requirement is loose enough that efficiency is preserved also for events decaying through the $\tau \to \rho\, \nu_\tau \to K\,\pi^0\,\nu_\tau$ and $\tau \to a_1\, \nu_\tau \to \pi\,\pi^0\,\pi^0\,\nu_\tau$ channels.

In events with two strips, a π^0 candidate is reconstructed combining the total momenta of the two strips, assuming each of them to be massless; the reconstructed π^0 mass is required to be in the $50\text{--}200\,\mathrm{MeV}/c^2$ range. The reconstruction then proceeds as for the single strip case.

Approximately 50% of the hadronic tau decays are reconstructed in the modes with one or two strips, more than 90% of which arise from genuine tau decays with one or more π^0.

In the final state topology with three charged hadrons, the reconstructed hadron tracks are required to be consistent with a common secondary vertex; a reconstr uction of the decay channel $\tau \to a_1\, \nu_\tau \to 3\pi\, \nu_\tau$ is attempted, requiring the reconstructed a_1 mass to be in the $0.8\text{--}1.5\,\mathrm{GeV}/c^2$ range. This last topology amounts for the remaining 25% of the reconstructed taus, of which about 90% are from three-hadron decays.

Once the particles corresponding to the reconstructed tau decay mode are identified, isolation criteria are applied on the remaining charged hadrons and photons in the jet. Three working points are defined, corresponding approximatively to misidentification probabilities for fake jets of 1% (loose), 0.5% (medium) and 0.25% (tight). Misidentification probabilities have been measured in data in different control samples of muon-enriched multijet and $W \to \mu\, \nu$ events, and found to be in agreement with the predictions from simulation to a level significantly better than the difference between the measurements in the two event samples. Reconstruction efficiencies and misidentification probabilities are shown in Figure 2.26.

Efficiency measurements on data.   The measurement of the reconstruction efficiency for hadronic taus from data is more challenging than the equivalent measurements for electrons and muons, because of the lower purity in the samples involved, and the worse ditau mass resolution because of the undetected neutrinos. A measurement has been performed from a simultaneous fit of reconstructed $Z \to \tau^+\tau^-$ events in different

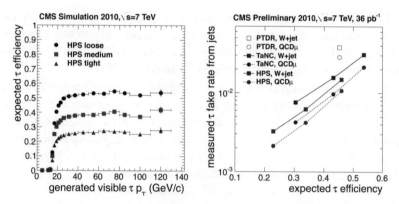

Figure 2.26. Left: tau identification efficiency for the three working points of the HPS algorithm, determined from simulated $Z \to \tau^+\tau^-$ events. Right: misidentification probabilities measured in data as function of expected efficiency from simulations, for different tau identification algorithms and working points. TaNC denotes another reconstruction particle-flow based reconstruction algorithm, while PTDR refers to the traditional fixed cone algorithms from TDR [55]

tau decay final states ($\mu\tau_h$, $e\tau_h$, $e\,\mu$, $\mu\,\mu$); the estimated ratio between reconstruction efficiencies in data and simulation with this technique is $(0.94 \pm 0.09)\%$ [56]. A cross-check of this measurement has been performed using the tag-and-probe method on $Z \to \tau^+\tau^-$ events with one tau decaying into a muon and the other decaying hadronically, obtaining compatible results within the 20–30% uncertainty [53].

Chapter 3
Search for a Higgs boson in the $H \to WW \to 2\ell2\nu$ channel

In the mass range from about 125 to 200 GeV/c^2, a very good sensitivity to a SM Higgs boson signal at the LHC is achieved in the $H \to WW \to 2\ell2\nu$ decay channel, thanks to the combination of the large branching fraction and the clean dileptonic final state with little QCD-induced background.

The irreducible background for this search is the non-resonant electroweak WW production. Due to the presence of two undetected neutrinos in the final state, the invariant mass of the system cannot be fully reconstructed; a good discriminating power can be nonetheless achieved exploiting the different kinematics of the processes.

The main reducible backgrounds in the analysis are W+jets, in case of lepton misidentification, and Drell-Yan, in case of instrumental missing transverse energy. The $t\bar{t}$ and associated tW production are also very important when considering final states with one or more jets, *e.g.* for vector boson fusion qq $\to$ H + 2jets, as they are irreducible except for the flavour content of the jets.

Minor reducible backgrounds, contributing only at the percent level to the overall yield, include: W + γ production followed by a $\gamma \to e^+e^-$ conversion in the detector material, if one of the two electrons is misidentified as a primary electron; W+γ^* and W+Z production with one lepton from the γ^*/Z escaping detection, and to an even lesser extent ZZ $\to 4\ell$ with two undetected leptons; Z $\to \tau^+\tau^-$ with both taus decaying leptonically; ZZ $\to 2\ell2\nu$.

The analysis strategy used at CMS is the following: first, events are categorized according to the number of hadronic jets in the final state, and separated in same flavour (ee, $\mu\mu$) and opposite flavour (e μ) events, to account for different background compositions; then the reducible backgrounds are strongly suppressed using a selection independent of the Higgs boson mass hypothesis, thereafter called WW selection; finally, the Higgs boson signal is extracted using the output distributions of multivariate classifiers trained individually for each Higgs boson mass hypothesis.

3.1. Triggers

For the H → WW analysis, events have been selected using single and double lepton triggers. In the 2011 LHC run, the thresholds and selection requirements on the trigger evolved with time to cope with the rising instantaneous luminosity and pile-up; to benefit from the lower thresholds available in the earlier part of the run, in each data taking interval the loosest available triggers have been used.

In addition to the main triggers used to select the signal events, other utility triggers have been used to collect the samples used to measure signal efficiencies and background rejection factors.

Muon triggers. Single muon triggers with and without isolation requirements have been used throughout all the 2011 run, with thresholds evolving from 15 to 40 GeV/c, and a pseudorapidity coverage up to $|\eta| = 2.4$, or $|\eta| = 2.1$ in the second half of the data taking (Section 2.4.5).

Double muon triggers with thresholds of $7/7$, $13/8$ and $17/8$ GeV/c have also been used, with no isolation requirements and pseudorapidity coverage always up to $|\eta| = 2.4$. In the last part of the run tracker muon reconstruction at trigger level was also exploited.

The average trigger efficiency from the combination of single and double muon triggers for signal events satisfying the offline selection requirement $p_T > 20/15$ GeV/c is about 97% for a Higgs boson mass of 120 GeV/c^2, and increases for higher masses.

Single muon triggers have also been used to collect the unbiased Z → $\mu^+\mu^-$ control samples to measure muon identification and trigger efficiencies. Control samples to measure the background rejection factors have been collected using pre-scaled single muon triggers with no isolation requirements and thresholds in the 9–24 GeV/c range.

Electron triggers. The thresholds for single electron triggers have been rapidly increasing from 27 GeV/c at the start of 2011 data taking to 80 GeV/c at the end. Therefore, the majority of signal events in the dielectron final state have been collected using double electron triggers, with p_T thresholds of $17/8$ GeV/c, and loose isolation and identification requirements (Section 2.5.6).

The average per-event trigger efficiency in the double-electron channel is about 98% for a light Higgs boson, and increases for higher masses.

In order to collect Z → e^+e^- events for measurements of the electron identification and trigger efficiencies, special double electron triggers have been used, with very tight requirements on one electron leg and minimal requirements on the other. Control samples to measure the background rejection factors have been collected using pre-scaled single electron triggers with p_T thresholds of 8–32 GeV/c and the same loose isolation and identification requirements of the main double electron trigger.

Cross-lepton triggers. Opposite flavour events have been collected using electron-muon triggers with p_T thresholds of 17 GeV/c and 8 GeV/c for the leading and trailing leptons respectively, independently of the flavour. Additional requirements on the electron part of the trigger were a loose identification in the early data taking, and a tighter identification combined with a loose isolation later.

Single muon and single electron triggers were also used to complement the cross-triggers, just like in the ee and $\mu\,\mu$ case. The average per-trigger efficiency in electron-muon events is close to that of double electron events.

3.2. Simulated datasets

Simulations of signals and background processes used throughout the analysis have been generated with different Monte Carlo generators:

- POWHEG [57] has been used for the signal samples [58, 59], Drell-Yan for events up to one hadronic jet [60], $t\bar{t}$ and single top processes [61, 62]. For all these processes, POWHEG provides a description of the event kinematic accurate up to next-to-leading-order QCD corrections, in addition to the soft-or-collinear radiation included in the parton shower Monte Carlos like PYTHIA.

- MADGRAPH [63] has been used for $q\bar{q} \rightarrow$ WW, W + jets, WZ, Wγ, Wγ^* and Drell-Yan for events with two or more jets. It has also been used to cross-check the description of the Drell-Yan in events with up to one jet. MADGRAPHis a matched leading-order Monte Carlo that includes the correct tree level matrix elements for final states with multiple hadronic jets.

- GG2WW [64] has been used for the gg $\rightarrow$ WW process. While this process is technically a second order correction to $q\bar{q} \rightarrow$ WW, it gives an $O(5\%)$ contribution to final yields due the large partonic luminosity for gluons and a kinematic more similar to that of H $\rightarrow$ WW.

- PYTHIA [65] has been used for inclusive ZZ events, and for the QCD multi-jet events used to study background rejections factors.

- MC@NLO [66] has been used as a cross-check of POWHEG for the signal and $q\bar{q} \rightarrow$ WW processes, and to evaluate the theoretical uncertainties related to missing higher orders. MC@NLO is a next-to-leading-order event generator similar to POWHEG, but relies on different theoretical computations for these physical processes, and accounts differently for the virtual QCD corrections.

Parton showering, underlying event and hadronization have been simulated with PYTHIA, except for the MC@NLO samples where HERWIG [67] has been used to assess the systematical uncertainties from the modelling of these effects; the TAUOLA [68] package has been used for tau decays.

All Monte Carlo samples have been processed through the full simulation of the CMS detector based on GEANT 4 [69], and reconstructed using the same algorithms used for the data; uncertainties in the detector alignment and calibration are included in the simulation. Simulation of in-time and out-of-time interactions is also included, and events are weighted so that the instantaneous luminosity profile of the simulation matches the one of the data. Simulation of the high level trigger has not been used in this analysis; the events are instead weighted proportionally to the trigger efficiency extracted from measurements in data on $Z \rightarrow \ell^+\ell^-$ events, as described later in Section 3.6.

Simulated Higgs boson events generated through the gluon fusion production mode have been weighted with a k-factor dependent on the Higgs boson p_T to match the NNLL+NLO computation from HQT. A similar procedure has been applied to Drell-Yan samples, using k-factors depending on the dilepton mass and rapidity computed with the FEWZ code [70].

3.3. Physics object reconstruction

The H $\rightarrow$ WW analysis is characterized by a large signal branching fraction but also large reducible backgrounds, so that choices related to physics object definitions are mostly driven by the requirement to strongly suppress those backgrounds.

Muons. On top of the tight muon selection described in Section 2.4.3, also stringent requirements on the transverse impact parameter of the muons are applied: $d_{xy} < 200$ μm for p_T above $20\,\text{GeV}/c$, $d_{xy} < 100$ μm below. The longitudinal impact parameter is also required to be less than 1 mm, mostly to suppress events in which one of the two leptons is not from the main interaction.

In the 2011 analysis, muon isolation is defined using the objects reconstructed with the particle flow algorithm in a cone of radius $R = 0.3$ around the muon; all charged particles originating within $\Delta z = 1$ mm from primary vertex are considered, while a p_T threshold of 1 GeV is applied to the neutrals ones to mitigate the effects of pile-up. The muon is accepted if the sum of the p_T of all the selected particles, excluding the muon itself, does not exceed a specified fraction of the muon p_T. The fraction used for muons in the barrel[1] are 0.13 or 0.06 for muons with p_T above or below 20 GeV/c, respectively; in the endcaps, where the purity before isolation is lower, tighter fractions of 0.09 and 0.05 are used for high and low p_T muons respectively.

In 2012, a multivariate isolation discriminator is used, to exploit also the angular distribution of the particle flow objects within the isolation cone. This provides a substantially better discrimination in the low p_T region, as shown in Figure 3.1.

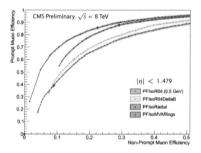 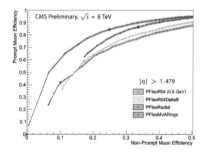

Figure 3.1. Performance comparison of different muon isolation criteria for muons with $p_T < 20$ GeV/c in the barrel (left) and endcaps (right). PFIsoR04 and PFIso04DeltaB refer to the traditional cone based isolation, with different pile-up correction schemes, while PFIsoRadial and PFIsoMVARings are approaches using also the angular distribution of particles in the isolation cone, either in a simplified analytical way or using an MVA. The latter approach is the one eventually used in the H $\rightarrow$ WW $\rightarrow$ $2\ell2\nu$ analysis.

Electrons. The starting point of the electron definition is a loose selection designed to reproduce the one applied at HLT, and the impact parameter requirements $d_z < 1$ mm and $d_{xy} < 200$ µm; a further selection is then applied using a multivariate classifier combining electron identification and impact parameter variables.

Electron isolation is based on particle flow objects, like for muons, but using a larger cone radius $R = 0.4$ and excluding from the computation the neutral hadrons close to the electron ($R < 0.07$, to account for possible leakage), and any electron and photon candidates in a ϕ strip

---

[1] In order to use the same isolation definition for electrons and muons, in this context the barrel region was defined to be $|\eta| < 1.479$ also for muons.

$\Delta\eta < 0.025$ (in order to exclude bremsstrahlung photons, possibly converted into e^+e^- pairs). The ratio between the sum of the p_T of the selected particles in the cone and the electron p_T is required to be less than 0.13 in the barrel and 0.09 in the endcaps.

Similarly to the muon case, in the 2012 analysis a multivariate isolation discriminator is further used to improve the rejection of the reducible background compared to the one achievable with the simple sum of the p_T of the particle flow candidates in the isolation cone.

In order to reject electrons from converted photons in the detector material, electron candidates are required to have a reconstructed hit in the innermost active tracker layer crossed. In addition, electrons are vetoed whenever a second track is found compatible with a conversion vertex; the compatibility is evaluated from the χ^2 of a constrained vertex fit using the two tracks.

Jets. Jets have been reconstructed from all particle flow candidates using the anti-k_T algorithm with distance parameter $R = 0.5$, as described in Section 2.6.1. Traditional jet energy calibrations as function of E_T and η are complemented by energy corrections for pile-up, based on the median energy density computed with the FASTJET algorithm [40]. For the purpose of event classification, jets with $p_T > 30\,\text{GeV}/c$ and $|\eta| < 5$ are considered; jets with smaller transverse momenta are still used in other selection requirements.

In this analysis, leptons are not removed from the particle flow candidates prior to the jet clustering, so leptons are also reconstructed as jets. This ambiguity is resolved by discarding jets within $R = 0.3$ of an identified and isolated lepton.

Heavy flavour tagging. An important tool to reject the backgrounds from top quark events is the identification of jets from bottom quarks. For this purpose, the track counting high efficiency algorithm is used (Section 2.3.6), applied to all jets with $p_T > 10\,\text{GeV}/c$.

In order to suppress the contribution from pile-up, jets are discarded if the longitudinal impact parameter of the jet with respect to the selected event primary vertex is larger than 2 cm; for this purpose, the jet impact parameter is defined as the p_T^2-weighted average of the impact parameters of the tracks within the jets.

The tagging algorithm based on track impact parameters is complemented by one relying on soft muons from semileptonic decays of heavy flavour hadrons. For this purpose, muons with p_T down to $3\,\text{GeV}/c$ are considered, with a specific selection: muons are required to be reconstructed as tracker muons, with tight track-to-segment matching criteria, and to have at least 11 hits in the inner tracker; in this context, the transverse and longitudinal impact parameter requirements are relaxed to 2 mm. Any muon from the candidate WW $\rightarrow$ $2\ell 2\,v$ decay is excluded from the soft muon list, and in order to strengthen the complementarity between this muon definition and the nominal one used for muons from W decays, in this context muons with $p_T > 20\,\text{GeV}/c$ are required to be anti-isolated, *i.e.* the total transverse momentum of all reconstructed particles in the isolation cone must be larger than 10% of the muon p_T.

Projected missing transverse energy.   The missing transverse energy E_T^{miss} is computed from the particle flow algorithm, using all reconstructed particles in the event. A projected missing energy scalar is introduced, defined as the component of the $\vec{E}_T^{\text{miss}}$ vector transverse to the direction of the nearest lepton in the azimuthal plane, if a lepton is found within $\Delta\phi < 90°$ from the $\vec{E}_T^{\text{miss}}$ vector; if both leptons are at $\Delta\phi > 90°$ from the missing transverse energy vector, the projected missing energy is defined as the full E_T^{miss}. This variable provides a better discrimination against the Z $\rightarrow$ $\tau^+\tau^-$ events, in which the missing transverse energy is preferentially aligned with leptons, and against those Z $\rightarrow$ $\ell^+\ell^-$ events in which the imbalance arises from poorly reconstructed leptons (Figure 3.2).

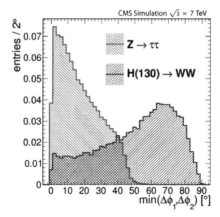

Figure 3.2. Angular separation between the $\vec{E}_T^{\text{miss}}$ vector and the closest lepton, for simulated Z $\rightarrow$ $\tau^+\tau^-$ and H $\rightarrow$ WW $\rightarrow$ $2\ell 2\,v$ events.

Minimum missing energy.   In order to improve the rejection of Drell-Yan, an additional vector $\vec{E}_{T,\text{trk}}^{\text{miss}}$ is computed using only the reconstructed charged particles associated to the selected event primary vertex, to avoid contamination from pile-up. For Drell-Yan events, with no genuine E_T^{miss} from neutrinos, the two variables E_T^{miss}, $E_{T,\text{trk}}^{\text{miss}}$ are found to be mostly uncorrelated, so that using both can provide a more effective suppression of this background (Figure 3.3).

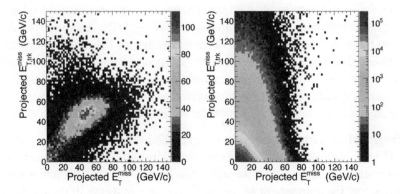

Figure 3.3. Joint distribution of the projected E_T^{miss} and $E_{T,\text{trk}}^{\text{miss}}$ on simulated signal events for a Higgs boson mass of 130 GeV/c^2(left) and simulated Drell-Yan events (right).

3.4. WW event selection

Leptons. Events are selected requiring the presence of two identified and isolated leptons, of opposite electrical charge. The leading lepton must have $p_T > 20$ GeV/c, and the trailing lepton $p_T > 10$ GeV/c (for e μ events) or $p_T > 15$ GeV/c (for ee, μ μ events) The invariant mass of the dilepton system $m_{\ell\ell}$ is required to be above 12 GeV/c^2 for opposite-flavour events, mostly to suppress correlated dileptons from hadron decays, and above 20 GeV/c^2 for same-flavour events. The higher p_T and $m_{\ell\ell}$ thresholds used for same-flavour events are useful to reduce the contamination from low mass $\gamma^* \rightarrow \ell^+\ell^-$ events. To suppress the Z + jets background, same-flavour events are rejected if the invariant mass of the two leptons is within 15 GeV/c^2 from that of an on-shell Z boson. The transverse momentum of the dilepton system is also required to be larger than 45 GeV/c, since the kinematic of the H $\rightarrow$ WW $\rightarrow 2\ell2\nu$ decay favours events in which the dilepton system is collimated and back-to-back to the neutrinos, while most background processes favour back-to-back leptons.

Missing energy requirements. In this search, a tight requirement on the missing transverse energy is used to suppress the Z/γ^* background. In the analysis of the 2011 data, the minimum of the projected missing transverse energies computed from all particle flow candidates E_T^{miss} and only primary charged particles $E_{T,\text{trk}}^{\text{miss}}$ is applied. is required to be larger than 20 GeV in opposite-flavour events. For same-flavour events, the threshold applied is higher, and dependent on the number reconstructed primary vertices as $(37 + N_{\text{vtx}}/2)$ GeV; this provides a better control of the Drell-Yan contamination in higher multiplicity events, which have a larger instrumental E_T^{miss} (Figure 3.4)

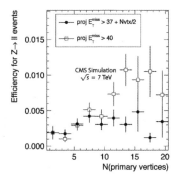

Figure 3.4. Efficiency of the minimum projected E_T^{miss} requirement for simulated $Z \rightarrow \ell^+\ell^-$ events as function of the number of reconstructed primary vertices, for the variable threshold used in the analysis $37 + N_{\text{vtx}}/2\,\text{GeV}$ (black), and for a fixed threshold of $40\,\text{GeV}$ (red).

The strategy for the same-flavour events has been revised in the analysis of the 2012, to deal with the larger pile-up: in the searches for a light Higgs boson ($m_H \leq 140\,\text{GeV}/c^2$) a dedicated multivariate discriminator based on missing transverse momentum and other kinematic and topological variables is used to suppress the Z/γ^* background, while for heavier Higgs boson hypotheses a fixed threshold of $45\,\text{GeV}$ is applied to the minimum of the two projected missing transverse energies. In the two-jet final state, where the substantial contribution from neutral particles spoils the performance of the $E_{T,\text{trk}}^{\text{miss}}$ variable and the $Z \rightarrow \tau^+\tau^-$ background is subdominant, the simple requirement $E_T^{\text{miss}} > 45\,\text{GeV}$ was found to be optimal.

Mismeasurement of jet energies is another potential source of instrumental E_T^{miss} in $Z + \text{jets}$ events. To suppress this component of the background, events are rejected if the azimuthal angle between the momenta of the dilepton system and the leading jet in the event is larger than $165°$, when the leading jet p_T is larger than $15\,\text{GeV}/c$ (Figure 3.5). In events with two or more jets with $p_T > 30\,\text{GeV}/c$, the requirement is applied using the momentum of the dijet system instead of the momentum of the leading jet.

Top quark and diboson rejection.   After the requirements on the leptons and missing transverse energy have been applied, the contamination from $W/Z + \text{jets}$ events is vastly reduced. To suppress the remaining backgrounds from WZ, $t\bar{t}$ and tW, events are rejected if there is a third lepton satisfying the nominal identification and isolation requirements, or

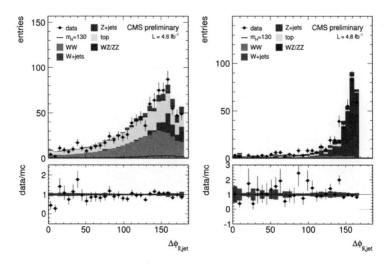

Figure 3.5. Distribution of the angular separation in the azimuthal plane between the dilepton system and the leading jet, in events with one jet. The distribution in the left panel is for events in the signal region, while the distribution in the right panel is from a control region enriched in Z/γ^* events. In both cases, the requirement $\Delta\phi(\ell\ell, \text{jet}) < 165°$ is applied in the same-flavour events.

a b-tagged jet, or a soft muon; the minimum p_T requirements for these objects are $10\,\text{GeV}/c$ for leptons and b-jets, and $3\,\text{GeV}/c$ for soft muons.

Results.   In the data collected in the 2011 LHC run, corresponding to an integrated luminosity of about $4.6\,\text{fb}^{-1}$, a total of 3071 are observed after the WW selection. A good agreement is found between the observation and the expectations from simulations, in terms both of yields (Table 3.1) and of distributions of the main kinematic variables (Figure 3.6, 3.7).

	0-jet	1-jet	2-jet
observed events	1359	909	703
expected bkg.	1365	952	715
expected sig.	70	29	16
WW	1040	441	160
$t\bar{t} + tW$	147	335	413
$W + \text{jets}/\gamma^{(\star)}$	126	87	49
$Z + \text{jets}$	19	60	78
$WZ + ZZ$	33	29	15

Table 3.1. Observed events in 2011 data after the WW selection, and expected backgrounds from the various physical processes, separately for events with zero, one, and two or more jets. The signal expectations refer to a SM Higgs boson of mass $130\,\text{GeV}/c^2$.

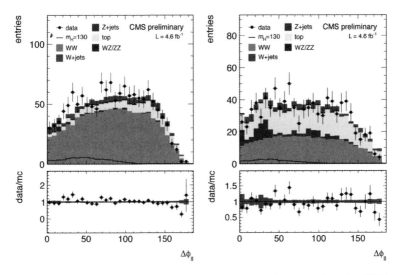

Figure 3.6. Azimuthal separation between the two leptons after the WW selection, for events with no jets (right) or one jet (left). The ratio between the observed and expected number of events in each bin is shown in the lower panel of each plot, the shaded grey are area representing the statistical uncertainty on the expected number of events.

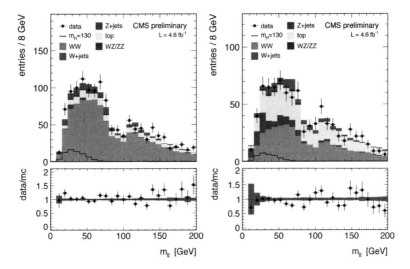

Figure 3.7. Invariant mass of the dilepton system after the WW selection, for events with no jets (right) or one jet (left). The deficit around $90 \, \mathrm{GeV}/c^2$ for both data and simulation arises from the Z veto applied only for same-flavour events.

A similar level of agreement is also observed in the 2012 data. At this level of the selection, the WW and $t\bar{t}$ processes are still dominating the total yield, so that the further steps are needed in order to infer about the presence or absence of a signal.

3.5. Signal extraction

While the presence of two undetected neutrinos in the H $\to$ WW $\to$ $2\ell 2\,\nu$ decay prevents the invariant mass of the system to be reconstructed, some discrimination against the irreducible WW background is achieved through the difference in the kinematic distributions between the two decays.

For Higgs boson masses below or close to the kinematic threshold $2m_W$, the two W bosons are produced at rest in the center-of-mass frame; as the Higgs boson is a scalar, angular momentum conservation requires the spins of the W^+ and W^- to be anti-parallel, and the chiral structure of the electroweak couplings causes the two leptons to be collimated, and back-to-back with the two neutrinos [71]. This peculiar topology provides a good discrimination against the continuum WW production.

For heavier Higgs bosons, instead, the momenta of the two W bosons becomes large, and partially overcomes the spin-alignment effect, reducing the discriminating power. Eventually, for large m_H the kinematics turns over completely and the two leptons are mostly back-to-back. The transition between the two regimes happens for Higgs boson masses around $225\,\text{GeV}/c^2$ (Figure 3.8). In addition to topological considera-

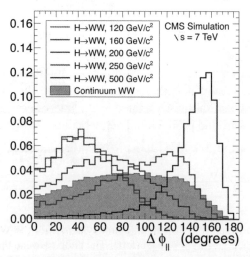

Figure 3.8. Angular separation between the two leptons in the transverse plane for the continuum WW background (filled histogram) and five different Higgs boson signal hypotheses (empty histograms). All distributions are normalized to the same area.

tions, estimators of the invariant mass of the system can be build from the lepton momenta and the missing transverse energy. In this analysis the transverse mass of the dilepton and E_T^{miss} system is used, defined as

$$m_T = \sqrt{2\, p_T^{\ell\ell}\, E_T^{\text{miss}}\, (1 - \cos \Delta\phi(\ell\ell, E_T^{\text{miss}}))}\,. \tag{3.1}$$

Other variables have been considered, and found to yield similar performances.

Different approaches to the signal extraction have been considered. The simplest one is a selection based on applying requirements to each kinematic variable individually; this approach is used in the analysis of the 2012 data, and of the two-jet category for the 2011 data; it is also used as a cross-check of the more advanced approaches in the other categories of the 2011 data.

Multivariate Analysis.   In the zero-jet and one-jet categories of 2011 data, a multivariate analysis of all the kinematics is performed: a boosted decision tree classifier (BDT) [72] is used to project the whole phase space into a single discriminating variable; inference about the signal is then performed by comparing the the distribution of the BDT output in data with the ones expected for the signal and background processes.

A loose selection dependent on the Higgs boson mass hypothesis is applied prior to the multivariate analysis, to improve the signal purity and to isolate away phase space regions that are used for background estimates: for light Higgs boson masses, the dilepton invariant mass $m_{\ell\ell}$ is required to be below a m_H-dependent threshold as defined in Table 3.2; the transverse mass m_T is required to be between $80 \,\text{GeV}/c^2$ and m_H, where the lower bound is mostly to suppress the $Z \to \tau^+\tau^-$ and $W + \gamma^{(*)}$ backgrounds. The events with $m_T > m_H$, potentially useful to normalize the WW background, are nonetheless excluded from the multivariate analysis to avoid potential biases related to the interference effects between signal and background production [73], which is not included in the simulations used.

The BDT classifier is trained on simulated events of $H \to WW$ signal and $q\bar{q} \to WW$ background, separately for the zero-jet and one-jet categories, using eight kinematic variables: the leading and trailing lepton transverse momenta $p_T^{\ell,\text{min}}$ and $p_T^{\ell,\text{max}}$; the dilepton invariant mass $m_{\ell\ell}$; the transverse mass of the system m_T, the separation between the two leptons in the azimuthal plane $\Delta\phi_{\ell\ell}$ and in the (η, ϕ) space $\Delta R_{\ell\ell} = \sqrt{\Delta\phi_{\ell\ell}^2 + \Delta\eta_{\ell\ell}^2}$; the two transverse masses obtained separately from each lepton and the E_T^{miss}. In the one-jet category, two additional variables are used: the azimuthal separation between the dilepton system

m_H (GeV/c^2)	$m_{\ell\ell}$ (GeV/c^2)	m_H (GeV/c^2)	$m_{\ell\ell}$ (GeV/c^2)	m_H (GeV/c^2)	$m_{\ell\ell}$ (GeV/c^2)
110	70	150	100	200	130
115	70	160	100	210	140
120	70	170	100	220	150
130	80	180	110	230	230
140	90	190	120	250	250

Table 3.2. Upper bound to the dilepton invariant mass $m_{\ell\ell}$ used in the selection prior to the BDT analysis, as a function of the Higgs boson mass m_H. For m_H of 230 GeV/c^2 and above, the kinematics of the decay changes and the requirement is simply $m_{\ell\ell} \leq m_H$.

and the E_T^{miss}, and between the dilepton system and the leading jet. An additional classification variable separating the different lepton flavours is also included in the training, since the different requirements applied in the WW selection result in a different kinematic.

Other multivariate approaches. Alternative approach to the boosted decision tree classifier have also been explored: the use of the distribution of a single kinematic variable, possibly after selections applied on the other variables; a factorized likelihood ratio obtained neglecting the correlations between the individual variables; a matrix element approach where the differential cross sections of signal and background are used directly [74]. The results with the different approaches were found to be similar, and the BDT was chosen for the nominal result because of the better expected sensitivity across the entire mass range.

VBF analysis. The search in the multi-jet topology is targeted to the vector boson fusion production mode. Additional selection requirements are therefore applied to the leading jets in the event: a large separation $|\Delta\eta(jj)| > 3.5$, and a large invariant mass $m_{jj} > 450\,\text{GeV}/c^2$; as the signal process is characterized by no colour connection between the two outgoing quarks, events are vetoed if any jet with $p_T > 30\,\text{GeV}/c$ are is found in the pseudorapidity region between the two jets.

Only one selection dependent on the Higgs boson mass is applied in the VBF case: for m_H below 200 GeV/c^2, the dilepton invariant mass is required to be below 100 GeV/c^2, while no requirement is applied for heavier Higgs boson masses. Except for that, the different sensitivities of the search for at the various m_H hypotheses derive only from the different production cross sections, branching ratios and selection efficiencies.

3.6. Signal modelling and systematical uncertainties

The predictions for the signal yields and kinematics are taken from Monte Carlo simulations, as described in Section 3.2. Systematical uncertainties of both experimental and theoretical origin on the signal model have been considered, and are described below. Except for a few theoretical uncertainties that are specific to the signal process, most of these uncertainties are considered also for those backgrounds that are estimated from simulations.

3.6.1. Experimental uncertainties

Efficiencies.   In order to correct for residual differences between the detector performance in data and simulations, lepton selection efficiencies are measured for $Z \to \ell^+\ell^-$ events in data and simulation as function of the lepton p_T and η. The differences are propagated to the expected signal yields by weighting events using the ratio of the two efficiencies; correction factors are found to be close to one within about 10%.

Simulation of the trigger is not used for this analysis. Instead, per-lepton trigger efficiencies are measured in data using $Z \to \ell^+\ell^-$ events as function of the lepton kinematic and the running condition, and each simulated signal event is assigned a weight equal to the probability for that event to pass the trigger. When considering both single lepton triggers and asymmetric double-lepton triggers, for each running period three per-lepton efficiencies are defined $\epsilon_S, \epsilon_{DH}, \epsilon_{DL}$, the last two referring to the high and low p_T legs of the double trigger. For the considered triggers, the selection applied to single leptons is strictly tighter than the one used in either leg of the double lepton trigger, and similarly the selection for the high p_T leg of the double lepton trigger is tighter or equal to that of the low p_T leg. Because of this, the correlation between the outcomes of the trigger selections is trivial, and the per-event efficiency can be decomposed as

$$
\begin{aligned}
\epsilon(1,2) = \quad & \epsilon_S(1) \times 1 & + \text{(from lepton 1 alone)} \\
& [\epsilon_{DH}(1) - \epsilon_S(1)] \times \epsilon_{DL}(2) & + \text{(add double 1+2 trigger)} \\
& [\epsilon_{DL}(1) - \epsilon_{DH}(1)] \times \epsilon_{DH}(2) & + \text{(add double 2+1 trigger)} \\
& [1 - \epsilon_{DL}(1)] \times \epsilon_S(2) & + \text{(add lepton 2 alone)}
\end{aligned}
$$

where (1) and (2) are taken to represent the variables of the first and second lepton in the event. The average efficiency on the full data-taking period is then obtained by averaging the per-event efficiencies of each running period with weights proportional to the integrated luminosities of each period. For this purpose, it is assumed that the per-lepton efficiencies for the e μ triggers are identical to the corresponding ones for the $\mu\,\mu$ and ee triggers; the validity of this assumption has been checked in data using a clean sample of dileptonic $t\bar{t}$ events collected using single-lepton triggers.

The uncertainties on the efficiency measurements from the tag-and-probe are dominated by the systematical uncertainties from the method rather than by the uncorrelated statistical uncertainties in each individual (p_T, η) region. The overall systematical uncertainty for the lepton selection and trigger is about 2% per lepton.

Energy scales.   The systematical uncertainties due to lepton energy scales and resolutions are evaluated by shifting or smearing the reconstructed lepton momenta in the simulation, and determining the corresponding changes in the selection efficiencies and in the output shape of the multivariate classifier used in the signal extraction. The typical impact of these uncertainties on the selection efficiencies is 1.5%, the precise value being dependent on the Higgs boson mass hypothesis considered. The same procedure is used to account for the systematical uncertainties from the jet energy scale calibration.

E_T^{miss}, **pile-up.**   While the pile-up degrades significantly the E_T^{miss} performance in $Z \to \ell^+\ell^-$ events, the effect is much smaller in the signal events which contain real E_T^{miss} from the two neutrinos. The systematical uncertainty on the signal efficiency and kinematics due to the possible mismodelling of the E_T^{miss} is evaluated for these events by varying the E_T^{miss} scale by 10%; the final impact on the selection efficiency is about 2%.

The uncertainty from the modelling of pile-up has been evaluated by varying the distribution of expected number of interactions per bunch crossing which is used to re-weight the simulation. The difference in the signal efficiency for a shift of the mean by ± 1 interaction was found to be 0.5%, negligible with respect to the other uncertainties involved.

Experimental uncertainties related to the jet counting procedure have been estimated comparing the jet selection efficiencies on $Z \to \ell^+\ell^-$ events in data and simulations; good agreement was found, and the resulting uncertainty was found to be negligible with respect to the theoretical ones.

Luminosity scale.   A 4.5% systematical uncertainty on the determination of the absolute scale of the integrated luminosity is assumed [75].

3.6.2. Theoretical uncertainties

Jet counting.   The theoretical uncertainties on the inclusive cross section for the different Higgs production mechanisms are taken from the latest theoretical predictions collected in reference [10]. However, as the analysis is performed in exclusive jet counting categories, a more complex treatment of the theoretical uncertainties is needed for the gluon fusion production, due to the sizable fraction of events with extra jets from initial state radiation.

A consistent prescription for handling the uncertainties related to jet counting has been developed in the context of the LHC Higgs cross section working group, and is summarised in references [76, 11]. From the theoretical point of view, the better defined quantities are the inclusive cross sections for the production of a Higgs boson plus at least N jets, $\sigma_{\geq N}$; the total production cross section σ_{tot} is then $\sigma_{\geq 0}$, and exclusive N-jet cross sections can be obtained as differences of inclusive ones, $\sigma_N = \sigma_{\geq N} - \sigma_{\geq N+1}$.

In the analysis, the theoretical uncertainty on the total cross section σ_{tot} is taken from reference [10], while the theoretical uncertainties on $\sigma_{\geq 1}$ and $\sigma_{\geq 2}$ are estimated using MCFM [77], varying the renormalization and factorization scales by a factor two around the nominal value of $m_\text{H}/2$; the MCFM computation is performed at NLO in the large m_top approximation, and then rescaled by the ratio of the LO cross sections computed with and without this approximation.

The theoretical uncertainties on the inclusive cross sections are taken to be statistically independent. The covariance matrix for the exclusive cross has a characteristic band-diagonal form, as for each N the exclusive prediction for σ_N is positively correlated with $\sigma_{\geq N}$, negatively correlated with $\sigma_{\geq N+1}$, and independent from all other $\sigma_{\geq K}$ with $K \neq N, N+1$.

Lepton acceptance. The uncertainties on the acceptance for leptons and missing energy arising from the knowledge of the parton distribution functions and strong coupling constant α_S have been evaluated following the PDF4LHC prescription [78, 79, 80, 81, 82], taking the envelop of the uncertainty bands from the three sets MSTW2008, CT10, NNPDF.

The size of the theoretical uncertainties on the acceptance from the missing higher order terms in the perturbative expansion has also been estimated, in the assumption that the uncertainties contribute only through the Higgs boson p_T spectrum[2]: simulated signal events have been reweighted to match the different Higgs p_T spectra computed with HQT for different values of the factorization and renormalization within a factor 2 from the nominal ones. The effect was found to be negligible with respect to the uncertainties related to jet counting.

---

[2] This assumption is motivated by the absence of QCD final state radiation corrections to the outgoing leptons.

Heavy Higgs boson modelling.   The natural width of the SM Higgs boson becomes large for masses above about $300\,\text{GeV}/c^2$, resulting in complex distortions of the lineshape when the convolution with the parton density functions is performed. As the effect is not well modelled in the current Monte Carlo generators, an additional uncertainty on the signal yield was included, dependent on the Higgs boson mass according to the *ad interim* LHC Higgs Cross Section working group prescription [83]:

$$\frac{\Delta\sigma}{\sigma} = 150\% \times \left(\frac{m_\text{H}}{1\,\text{TeV}}\right)^3.$$

The resulting uncertainty is about 10% for a Higgs boson mass of $400\,\text{GeV}/c^2$, and about 30% at $600\,\text{GeV}/c^2$. An improved modelling of the effect is being developed in the LHC Higgs cross section working group.

Other phenomenological uncertainties.   The uncertainties arising from the modelling of the soft-or-collinear QCD radiation, the underlying event and the hardonization process have been evaluated by comparing the efficiencies extracted from two different Monte Carlo generators: PYTHIA [65], applied to events generated with POWHEG; and HERWIG [67], applied to events generated with MC@NLO, using the underlying event model provided by JIMMY [84]. For the zero and one jet final states, the resulting uncertainties are in the 5–13% range, anti-correlated as they arise mostly from event migration between the two categories. The uncertainty in the two-jet final state for the gluon fusion process is larger, up to 25% at low Higgs boson masses; the impact of this uncertainty on the overall sensitivity of the two-jet analysis is nonetheless small, since the signal yield in this topology is dominated by the vector boson fusion production mechanism.

3.7. Background predictions

The estimation of the backgrounds from other standard model processes is a crucial point in the H $\rightarrow$ WW search. A strong preference is given to the techniques to determine the backgrounds from the data, especially for reducible ones like W + jets and Z + jets, to minimize the potential biases from the modelling of the detector response. Information from Monte Carlo simulations is also used, both to extrapolate some background measurements from control regions in data to the signal region, and to estimate directly the yields of some minor backgrounds like WZ and ZZ.

3.7.1. W + jets and QCD multijet backgrounds

This class of reducible backgrounds is arises from the misidentification of leptons from heavy flavour decays, charged hadrons or electrons from photon conversions. To estimate these backgrounds from data, control samples are defined by the events in which one or both leptons satisfy only some loose identification requirements and not the ones applied in the nominal selection, and evaluating the probability that a loose lepton passes the tight selection.

The lepton misidentification probability is measured separately from a second control sample enriched in QCD di-jet events, and the two information are combined to obtain an estimate of the background events satisfying the nominal selection.

Loose lepton definitions. For electrons, the loose selection is defined to reproduce the identification and detector-based isolation requirements that are applied at trigger level, combined with the full conversion rejection used in the nominal selection and the impact parameter requirement $d_{xy} < 200$ µm; compared to the nominal selection, the multivariate electron identification and the particle-flow-based isolation requirements are dropped.

For muons, the loose selection is defined by relaxing partially the transverse impact parameter and the isolation requirements.

Misidentification probabilities. The misidentification probabilities, sometimes referred to as "fake rates" or "loose-to-tight ratios", are determined in a control sample selected requiring one loose lepton and at least one hadronic jet well separated from the lepton.

The contributions from W +jets and Z + jets events are suppressed by tight requirements on the missing transverse energy and on the transverse mass of the lepton-E_T^{miss} system, and by vetoing additional leptons. The remaining contamination after these requirements is estimated from simulation and subtracted. This contamination is at the percent level for low momentum leptons, but increases strongly with p_T to about 30% at 30 GeV/c, and about 50% at 40 GeV/c; as the misidentification probabilities are expected from simulations be approximately flat for transverse momenta above about 30 GeV/c, the probabilities for leptons with $p_T > 35$ GeV/c^2 assumed to be equal to those measured for $p_T \in [30, 35]$ GeV/c.

In order to evaluate empirically the systematical uncertainty on this estimation, the measurement is repeated varying the p_T requirements on the hadronic jet in a range comparable to the p_T spectrum of jets in W + jets events. A cross-check is also made by performing the measurement on γ + jet events, *i.e.* replacing the requirements of a hadronic jet with that of a photon; this results in a different jet flavour composition, since

$\gamma +$ jets is enriched in quark-induced jets and depleted in gluon-induced ones with respect to QCD multijet events.

As the misidentification probabilities are to some extent dependent on the pile-up through the isolation requirements, the measurement is also performed separating the events in categories according to the multiplicity of reconstructed primary vertices.

Measured values of the misidentification probabilities for electrons range from about 2% to 8% depending on the electron p_T and η; smaller values are for lower momentum electrons, especially in the endcaps, where the nominal selection is tighter. Misidentification probabilities for muons are larger, ranging from about 15% to 30%, since the requirements of the loose and tight selections are much closer than in the electron case. The misidentification probabilities for the barrel region are shown in Figure 3.9.

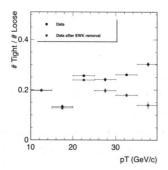 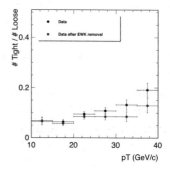

Figure 3.9. Misidentification probabilities in the barrel for muons (left) and electrons (right), before and after the subtraction of the contamination from W and Z (black and red dots respectively).

Background determination. The normalization of the W + jets and QCD multijet background can be determined through the relation between the observed number of events with zero, one or two loose leptons failing the nominal selection N_{0F}, N_{1F}, N_{2F}, the misidentification probabilities ϵ_B and the efficiencies of the nominal selection for signal leptons ϵ_S. This result in a linear system of three equations:

$$
\begin{aligned}
N_{2F} &= (1-\epsilon_S)^2 N_{\ell\ell} + (1-\epsilon_S)(1-\epsilon_B)N_{\text{W+jets}} + (1-\epsilon_B)^2 N_{\text{QCD}} \\
N_{1F} &= 2\epsilon_S(1-\epsilon_S)N_{\ell\ell} + (\epsilon_S+\epsilon_B-2\epsilon_S\epsilon_B)N_{\text{W+jets}} + 2\epsilon_B(1-\epsilon_B)N_{\text{QCD}} \\
N_{0F} &= \epsilon_S^2 N_{\ell\ell} + \epsilon_B\epsilon_S N_{\text{W+jets}} + \epsilon_B^2 N_{\text{QCD}}.
\end{aligned}
$$

$$(3.2)$$

The three unknowns in this system of equations are N_{QCD}, $N_{\text{W+jets}}$ and $N_{\ell\ell}$, the last one being overall event yields for all the processes with two signal-like leptons, all defined after the loose selection requirements. After solving the equations, the W + jets and QCD multijet background estimates for the nominal selection are given by $\epsilon_B \epsilon_S N_{\text{W+jets}}$ and $\epsilon_B^2 N_{\text{QCD}}$.

The solution of the system of equations (3.2) becomes intuitive in the approximation that $\epsilon_S = 1$,

$$
\begin{aligned}
N_{2F} &= 0\,N_{\ell\ell} + & 0\,N_{\text{W+jets}} + & (1 - \epsilon_B)^2\,N_{\text{QCD}} \\
N_{1F} &= 0\,N_{\ell\ell} + (1 - \epsilon_B)\,N_{\text{W+jets}} + & 2\epsilon_B(1 - \epsilon_B)\,N_{\text{QCD}} \\
N_{0F} &= 1\,N_{\ell\ell} + & \epsilon_B\,N_{\text{W+jets}} + & \epsilon_B^2\,N_{\text{QCD}},
\end{aligned}
$$

for which the resulting background yields after the full selection are

$$
\epsilon_B^2\,N_{\text{QCD}} = \frac{\epsilon_B^2}{(1 - \epsilon_B)^2}\,N_{2F}
$$

$$
\epsilon_B\,N_{\text{W+jets}} = \frac{\epsilon_B}{(1 - \epsilon_B)}\,N_{1F} - \frac{2\epsilon_B^2}{(1 - \epsilon_B)^2}\,N_{2F}.
$$

As in practice ϵ_B is small, the second term can be neglected and the background prediction can be obtained by scaling the N_{1F} yield by a factor $\epsilon_B/(1 - \epsilon_B)$. In this approximate procedure, the fact that ϵ_S is not equal to one can be then corrected subtracting from the background prediction a term proportional to $(1 - \epsilon_S)/\epsilon_S$ determined from simulation.

Background distributions. The background estimates obtained by solving of the system of equations (3.2), or the approximation described above, are a linear combination of the three event yields N_{0F}, N_{1F}, N_{2F} with coefficients depending on ϵ_B and ϵ_S. The coefficients can then be interpreted as weights associated to each individual event of the three samples $0F$, $1F$, $2F$; the dependency of ϵ_B and ϵ_S as function of the lepton kinematics and of the multiplicity of reconstructed vertices can be then taken into account naturally in the weights[3]. The distribution of any kinematic variable for the W + jets background, including the output of the multivariate discriminator used for the signal extraction, are then estimated from the distribution of the weighted events. The impact of the uncertainties on the misidentification probabilities on the distributions can also be assessed by varying the values of ϵ_B used to compute the weights.

Systematical uncertainties. The dominant uncertainty in the final background estimation arises from the systematical uncertainty on the

---

[3] Formally this procedure can be interpreted as partitioning the five-dimensional space defined by the η and p_T of the two leptons and the number of primary vertices in a large number of bins, solving the system of equations in each bin and then adding up the results.

misidentification efficiency, and amounts to $\sim 36\%$. The estimate has also been checked by comparing the outcome of this estimation procedure to simulated events with the direct evaluation from simulated W+jets events satisfying the full selection requirements; the level of agreement observed between the two estimates is about 30%, and is covered by the uncertainty itself.

Same-sign check.   A validation of the procedure can be done by applying it to the events where the two leptons have the same charge, populated mostly by W + jets events. Good agreement is observed between the data and the predictions for the W + jets obtained with this method (Figure 3.10); for this test, the remaining contributions from $W + \gamma^{(*)}$ and $W + Z$ are estimated using simulations, and amount to about one third of the event yield.

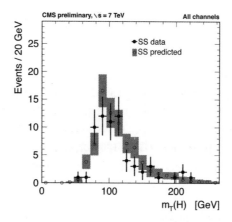

Figure 3.10. Transverse mass of the system, for same-sign events passing the WW selection in the zero-jet bin. The observation from data, shown as black points with error bars, is compared to the background prediction, shown as empty markers and grey error rectangles.

3.7.2. $W\gamma^{(*)}$ background

The $W + \gamma$ process can give rise to a final state similar to the signal if the photon converts into an electron pair in the detector material, one of the two electron is lost and the other electron is not identified as a conversion.

This contributions is not already included in the reducible W + jets background estimation from data, as the fraction of jet + γ events in the inclusive jet + X sample used to measure misidentification probabilities is much smaller than the corresponding fraction of W + γ events in the inclusive W + X sample to which the misidentification probabilities are applied. The inclusive $W + X$ background at the final selection is given by

$$
N_{\text{tight}}^{W+X} = \epsilon_{\text{tight}}^{W+\text{jets}} N_{\text{loose}}^{W+\text{jets}} + \epsilon_{\text{tight}}^{W+\gamma} N_{\text{loose}}^{W+\gamma} =
$$
$$
= \left(\epsilon_{\text{tight}}^{W+\text{jets}} \frac{N_{\text{loose}}^{W+\text{jets}}}{N_{\text{loose}}^{W+X}} + \epsilon_{\text{tight}}^{W+\gamma} \frac{N_{\text{loose}}^{W+\gamma}}{N_{\text{loose}}^{W+X}} \right) \times N_{\text{loose}}^{W+X} , \qquad (3.3)
$$

but the estimation is done with misidentification probabilities taken from inclusive jet $+ X$ events,

$$
N_{\text{estimate}}^{W+X} = \frac{N_{\text{tight}}^{\text{jet}+X}}{N_{\text{loose}}^{\text{jet}+X}} \times N_{\text{loose}}^{W+X}
$$
$$
= \left(\epsilon_{\text{tight}}^{\text{jet}+\text{jet}} \frac{N_{\text{loose}}^{\text{jet}+\text{jet}}}{N_{\text{loose}}^{\text{jet}+X}} + \epsilon_{\text{tight}}^{\text{jet}+\gamma} \frac{N_{\text{loose}}^{\text{jet}+\gamma}}{N_{\text{loose}}^{\text{jet}+X}} \right) \times N_{\text{loose}}^{W+X} .
$$

$$(3.4)$$

Since the fraction of jet$+\gamma$ events in the inclusive jet$+ X$ sample is much smaller than the corresponding fraction of $W + \gamma$ events in the inclusive $W+X$ sample, the right hand side of equation (3.4) is practically equal to $\epsilon_{\text{tight}}^{\text{jet}} N_{\text{loose}}^{W+X}$, which underestimates the $W + \gamma$ by a factor $\epsilon_{\text{tight}}^{\text{jet}}/\epsilon_{\text{tight}}^{\gamma} \ll 1$. Because of this, the $W + \gamma$ background yield and distributions are derived separately using simulated events, a conservative 30% uncertainty is assigned to this estimate.

A similar background topology arises from $W + \gamma^*$ production with $\gamma^* \to \ell\ell$, if one of the two leptons is not detected, except that final states with muons are also possible. This background is also estimated using simulations, and added to the $W + \gamma$ estimate. However, in this case the production cross section for $W + \gamma^*$ is normalized from data using a control sample of trilepton events from the decay $W + \gamma^* \to \ell\nu + \mu^+ \mu^-$ selected with the requirement $m_{\mu\mu} < 12\,\text{GeV}/c^2$ (Figure 3.11); a normalization factor of about 1.5 with respect to the leading order cross section from MADGRAPH is determined from this measurement[4].

As the background contributions from $W+\gamma^{(*)}$ have an equal chance of producing same sign or opposite sign leptons, the background estimates for these processes are also validated together with the ones for $W + $ jets in the control sample with same-sign leptons.

3.7.3. Z + jets background

Despite the much tighter requirements applied in the same-flavour channel, some contamination from the $Z/\gamma^* \to \ell^+\ell^-$ is still present. As this

---

[4] This is comparable with the scale of the k-factors with respect to leading cross sections found for similar electroweak processes.

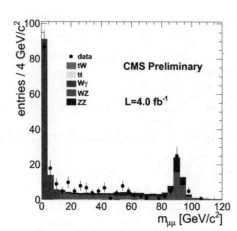

Figure 3.11.   Comparison of the dimuon invariant mass distribution in $\ell^{\pm}\,\mu^{+}\,\mu^{-}$ events in data and simulations. The contribution from $W + \gamma^{*} \to \ell\,\nu + \mu^{+}\,\mu^{-}$ decays is normalized from the $m_{\mu\mu} < 12\,\text{GeV}/c^2$ region.

background arises from instrumental issues related to pile-up, it is important to estimate it from the data, and a natural control region is given by the events with dilepton invariant mass $m_{\ell\ell}$ close to that of an on-shell Z boson.

Control region definition.   Separately for each jet multiplicity, a Z peak control region is defined by the requirement that $m_{\ell\ell}$ be within $7.5\,\text{GeV}/c^2$ from the nominal Z mass; all the requirements of the WW selection except the ones related to $m_{\ell\ell}$ are also applied. The events populating the control region can be separated in three categories: single boson events $Z \to \ell^{+}\ell^{-}$, resonant diboson events $(Z \to \ell^{+}\ell^{-}) + (W/Z \to \text{any})$, and all other non-resonant events.

The non-resonant contribution can be estimated from $e^{\pm}\,\mu^{\mp}$ selected with the same kinematic requirements. In order to account for the different efficiencies and acceptances for electrons and muons, correction factors are determined from the measured ratio of $Z \to e^{+}e^{-}$ and $Z \to \mu^{+}\mu^{-}$ yields without the E_T^{miss} requirement,

$$k_{e/\mu} = \frac{\epsilon(e)}{\epsilon(\mu)} = \sqrt{\frac{N_{Z \to e^{+}e^{-}}^{\text{no } E_T^{\text{miss}}}}{N_{Z \to \mu^{+}\mu^{-}}^{\text{no } E_T^{\text{miss}}}}} \qquad k_{e/\mu} = \frac{\epsilon(\mu)}{\epsilon(e)} = \frac{1}{k_{\mu/e}}, \qquad (3.5)$$

and the non-resonant background $N_{\text{peak}}^{\text{NR}}$ in the Z peak control region is estimated as

$$N_{\text{peak}}^{\text{NR}}(ee) = \frac{1}{2} k_{e/\mu}\, N_{\text{peak}}(e\mu) \qquad N_{\text{peak}}^{\text{NR}}(\mu\mu) = \frac{1}{2} k_{\mu/e}\, N_{\text{peak}}(e\mu). \qquad (3.6)$$

The resonant diboson background in the Z peak after the nominal selection is dominated by events with real E_T^{miss} from neutrinos, either from $Z \rightarrow \nu\nu$ decays or from $W \rightarrow \ell\nu$ decays where the lepton is not reconstructed or does not pass the selection requirements used for the third lepton veto. As the E_T^{miss} is not from instrumental effects, this contribution can be reliably predicted using simulated events; a 10% systematical uncertainty is assigned to this estimate.

The Z+jets yield determined from the Z peak observed in data is found to be about a factor three larger than the predictions from the simulation, roughly independent of the lepton flavour and jet multiplicity within the uncertainties on the estimate. The discrepancy is interpreted as a mis-modelling of the degradation of the E_T^{miss} performance in high pile-up events, as it decreases when the E_T^{miss} requirement is relaxed or if the comparison is restricted to events with a small number of reconstructed primary vertices. After the normalization of the $Z + \mathrm{jets}$ in the simulation is scaled to match this measurement, the kinematic distributions of the events in the peak are found to be well modelled by the simulation. Among the kinematic variables used as input to the BDT classifier, the only one where some distortion is visible is the reconstructed transverse mass of the Higgs boson, as it depends directly on the E_T^{miss} scale (Figure 3.12).

Extrapolation to the signal region.   The measured $Z + \mathrm{jets}$ yield in the Z peak control region is propagated to the signal region using the ratio of the expected yields in the simulation for those two regions:

$$
N_{\mathrm{signal}}^{Z+\mathrm{jets}} = \left(\left[N_{\mathrm{peak}} - N_{\mathrm{peak}}^{NR} \right]_{\mathrm{data}} - \left[N_{\mathrm{peak}}^{VV} \right]_{\mathrm{sim}} \right) \times \left[\frac{N_{\mathrm{signal}}^{Z+\mathrm{jets}}}{N_{\mathrm{peak}}^{Z+\mathrm{jets}}} \right]_{\mathrm{sim}} , \quad (3.7)
$$

where N_{peak}^{VV} is the yield in the Z peak from resonant diboson events, and the subscripts data and sim refer to quantities evaluate using data or simulated events respectively.

The nominal value of extrapolation factor is taken from simulated $Z +$ jets events. In order to reduce the statistical uncertainties from the limited number of simulated events, the requirement on the minimum of the projected E_T^{miss} and $E_{T,\mathrm{trk}}^{\mathrm{miss}}$ is relaxed from $(37 + N_{\mathrm{vtx}}/2)\,\mathrm{GeV}$ to $30\,\mathrm{GeV}$. The dependency of the extrapolation factor on the E_T^{miss} requirement is studied simulated events, and the largest deviation with respect to the nominal value is assigned as a systematical uncertainty on the factor.

For relaxed E_T^{miss} requirements, the $Z +$ jets background can also be estimated directly from the data in the signal region as it is done in the control region, *i.e.* by subtracting the non-resonant contributions using

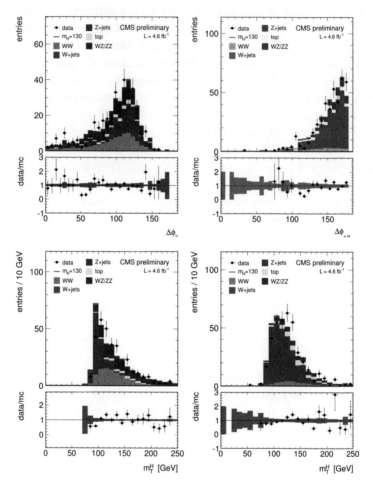

Figure 3.12. Some kinematic distributions of events in the Z peak after the WW selection: azimuthal separation between the two leptons (top left), azimuthal separation between the dilepton pair and the E_T^{miss} vector (top right), and transverse mass of the Higgs boson reconstructed from the dilepton-E_T^{miss} system. The plots in the left and right column are for the zero-jet and one-jet categories respectively.

opposite-flavour events and the dibosons contributions from simulation. This estimate is used to validate the extrapolation factor obtained from simulations; in general, a good agreement is observed.

Background distributions.   In addition to the normalization of the background, also an estimate of the output distribution of the multivariate classifier for Z + jets events is needed for the signal extraction. The nominal estimate is again taken from simulations relaxing the E_T^{miss} requirement in order to reduce the statistical fluctuations; the distribution observed in the simulation with the the nominal E_T^{miss} requirement is compatible with this estimate within the large statistical uncertainties. A partial validation of the kinematics has also been made comparing the output distributions of the multivariate classifier in data and simulations in events a control region with small E_T^{miss}, where the Z + jets process is dominant, and a satisfactory agreement is observed.

Systematical uncertainties.   The two dominant sources of uncertainty on the estimation of the Z + jets background are the uncertainty on the Z + jets event yield in the peak, of mostly statistical origin, and the systematical uncertainty on the extrapolation factor to the signal region. The overall uncertainty on the normalization of the background amounts to about 50%.

3.7.4. $Z \to \tau^+ \tau^-$ background

At first look, the $Z \to \tau^+ \tau^-$ process with both taus decaying leptonically could appear as a major background to the H $\to$ WW searches: it can contribute to both same-flavour and opposite-flavour states, it has real E_T^{miss} from neutrinos, and it has a dilepton invariant mass distribution concentrated at small $m_{\ell\ell}$, similar to that of a light Higgs boson.

However, the kinematic of $Z \to \tau\tau \to 2\ell\,4\,\nu$ decays is different enough from that of a H $\to$ WW $\to 2\ell\,2\,\nu$ that this process is strongly suppressed by the selection requirements used in the analysis.

First, the lepton momentum spectrum for this background is softer, since a sizable fraction of the energy is carried away by the neutrinos, so already the minimum p_T requirements applied to the two leptons reduces the yield significantly; likewise, the tight impact parameter requirements provide some level of rejection against leptons from tau decays.

Then, taus from the $Z \to \tau^+ \tau^-$ process are preferentially back-to-back and balanced in transverse momentum, and neutrinos from the tau decays are mostly collinear with the leptons: this results in a significantly smaller projected E_T^{miss} with respect to the signal; for the same reason, the background is strongly suppressed by the $p_T^{\ell\ell} \geq 45\,\text{GeV}/c$ requirement.

Already at the WW selection level, the fraction of background events arising from this process is a few percent. This contribution is then reduced to a negligible level by the further requirement $m_T > 80\,\text{GeV}/c^2$ applied before the signal extraction.

Estimation from tau embedding.   As the projected E_T^{miss} requirement is one of the dominant tools to suppress this background, it is important to assess that the rejection performance in data is not spoiled by the instrumental E_T^{miss} from pile-up. This is achieved using the tau embedding technique: a clean sample of $Z \to \mu^+ \mu^-$ is selected in data, and the two muons are replaced with simulated taus; as this method relies on the modelling of the detector response only for the leptons and neutrinos from the tau decay, the estimate is more robust than what could be achieved using simulated $Z \to \tau^+\tau^-$ events. Similarly to what observed for the $Z \to \ell^+\ell^-$ background, the simulation is found to underestimate the contribution from this process, as it does not reproduce well the degradation of the E_T^{miss} reconstruction performance with from pile-up.

3.7.5. Top quark background

The $t\bar{t}$ and tW processes contribute significantly to the background, especially in the final states with one or more jets.

The general procedure used to estimate this background is to measure the background yield in a control region defined inverting the b-jet veto, fully or partially, and then to extrapolate to the signal region using b-tagging efficiencies measured from data in a second control region; this second control region will be denoted as "efficiency measurement region" to avoid confusion between the two.

The implementation of this procedure is necessarily different in each jet multiplicity final state. As in the rest of the analysis, the categorization in jet multiplicities is done by counting the number of jets with $p_T > 30 \, \text{GeV}/c$; jets below this threshold will be denoted in the following as soft jets.

Zero-jet event category

Control region.   For the category of events with no jets with $p_T > 30 \, \text{GeV}/c$, the control region is determined by the events with one or more soft jet tagged as b-jet using the track counting high efficiency algorithm or by the presence of a soft muon. The contamination in this control region arising from processes other than $t\bar{t}$ or single top is subtracted using predictions from simulations; for this subtraction, the simulated W + jets and Z + jets events are scaled to to match the background estimates from data at the WW selection level described in the previous sections; likewise, the WW contamination is scaled by a factor 1.1 to match the larger b-jet mistag rate observed in data with respect to simulations (Section 2.3.6). The overall contamination in this control region is estimated to be about 20%, with a 40% relative uncertainty. A few kinematic distributions for the events in the control region are shown in the left panels of Figure 3.13; in general, good agreement is observed between data and predictions from simulations.

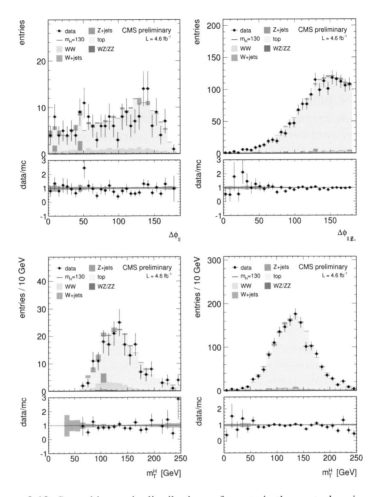

Figure 3.13. Some kinematic distributions of events in the control region used to measure the $t\bar{t}$ and tW backgrounds, after the WW selection: azimuthal separation between the two leptons (top left), azimuthal separation between the dilepton pair and the E_T^{miss} vector (top right), and transverse mass of the Higgs boson reconstructed from the dilepton-E_T^{miss} system. The plots in the left and right column are for the zero-jet and one-jet categories respectively.

Efficiency measurement. A control sample to extract the tagging efficiency measurement for soft jets is defined by the events with a single jet with $p_T > 30 \, \mathrm{GeV}/c$, tagged as b. This sample is mostly populated by $t\bar{t}$ events, with a sizable contribution from tW (about 25%); contaminations from non-top processes are at the percent level. The b-tagging efficiency per soft jet ϵ_{1b} is determined by the fraction of events in the control region that have a soft b-jet tag, after subtracting the expected yields for tW and the other non-top backgrounds from simulations,

$$\epsilon_{1b} = \frac{N_{\text{data}}(\text{tag}) - k_S\, k_H\, N_{\text{sim}}^{tW}(\text{tag}) - N_{\text{sim}}^{\text{other}}(\text{tag})}{N_{\text{data}}(\text{all}) - k_H\, N_{\text{sim}}^{tW}(\text{all}) - N_{\text{sim}}^{\text{other}}(\text{all})}, \tag{3.8}$$

where k_S and k_H are the ratio of b-tagging efficiencies in data and simulation for soft and hard jets; k_H is estimated as part of the top background determination in the one-jet category as described below, while k_S can be determined iteratively as the ratio of ϵ_{1b} in data and simulation from equation (3.8) itself. The impact of k_S and k_H on the determination of ϵ_{1b} is in any case small compared to the statistical uncertainties. The overall uncertainty on ϵ_{1b} measurement using this procedure is about 4%.

Extrapolation.   The average efficiency to tag a $t\bar{t}$ or tW event with no hard jets could be defined in terms of ϵ_{1b} as

$$\epsilon_{\text{top}} = f_{t\bar{t}} \times \left(2\,\epsilon_{1b} - \epsilon_{1b}^2\right) + f_{tW} \times \epsilon_{1b} \tag{3.9}$$

where $f_{t\bar{t}}$ and f_{tW} are the expected relative fractions of $t\bar{t}$ and tW at the WW selection level before the b-jet veto, estimated from simulations, and the combinatoric factor $\left(2\,\epsilon_{1b} - \epsilon_{1b}^2\right)$ is motivated by the two soft jets in $t\bar{t}$ events.

However, when applied to simulated $t\bar{t}$ and tW events events, the estimate using equations (3.8) and (3.9) reproduces the tagging efficiency for events with no jets with a bias of a few percent. To remove this bias, the definition is then modified as

$$\epsilon_{\text{top}} = (f_{t\bar{t}} + x\, f_{tW}) \times \left(2\,\epsilon_{1b} - \epsilon_{1b}^2\right) + (1 - x)\, f_{tW} \times \epsilon_{1b} \tag{3.10}$$

where the parameter x is chosen so that the procedure gives perfect closure when applied to simulated events.

The value of ϵ_{top} measured from data is then used to extrapolate from the control region of tagged events to the signal region as

$$N_{\text{signal}}^{\text{top}} = \left(N_{\text{tag}} - N_{\text{tag}}^{\text{other}}\right) \times \left[\frac{1 - \epsilon_{\text{top}}}{\epsilon_{\text{top}}}\right], \tag{3.11}$$

where $N_{\text{tag}}^{\text{other}}$ is the contamination in the control region. The overall uncertainty on the estimation is about 25%, including also the theoretical uncertainties on the prediction about the fraction of tW events and on the modelling of the tW process at NLO[5].

---

[5] The two processes tW and $t\bar{t}$ are no longer well defined separately at NLO, as they share some Feynmann diagrams [85]. Two recent prescriptions to define the tW process at NLO are the diagram removal and diagram subtraction [86], in which the double-resonant $t\bar{t}$ diagrams are removed from the tW process at amplitude level or subtracted later at cross section level. In this analysis, the former approach was used, but the difference between the two predictions has been taken as a systematical uncertainty.

One-jet event category

The procedure used to determine the top background in the category of events with a single jet above the p_T threshold of $30\,\mathrm{GeV}/c^2$ is simpler, as the control region can be defined by inverting the b-tag veto on the leading jet alone, preserving the veto on additional soft b-tagged jets applied in the WW selection. The purity of top events in this control region is about 95%, significantly higher than in the corresponding one for the zero-jet category; some kinematic distributions for the events in this control region are shown in the right panels of Figure 3.13.

A suitable control region to measure the b-tagging efficiency for the leading jet ϵ_b is obtained by selecting events with two jets and considering only the leading jet, whose kinematics is found to be more close to that of one-jet events; to improve the purity of the sample, it is further required that either the subleading jet or a soft jet in the event is b-tagged. This measurement is validated by comparing the ratio of efficiencies in data and simulation with respect to the ones obtained from other measurements of b-tagging efficiency, *e.g.* from QCD jet events (Section 2.3.6).

The propagation to the signal region is then performed as in equation (3.11), where in this case ϵ_{top} is equal to ϵ_b. The overall uncertainty on the estimate of the top quark background in the one-jet event category is about 6%; the significantly smaller uncertainty with respect to the zero-jet category is due to the much larger population and purity of the control sample, the higher precision with which the b-tagging efficiency can be determined at higher p_T, and the simpler extrapolation procedure.

Two-jet event category

For two-jet events, the control region is defined by considering the events in which the most central of the two jets is b-tagged; the full VBF selection is applied also in the control region, namely the $\Delta\eta_{jj}$ and m_{jj} requirements and the veto of other jets in the η interval between the two leading ones.

The b-tagging efficiency ϵ_b is measured from inclusive two jet events, *i.e.* without VBF selection, as for the one-jet category. However, since the kinematic is different, the tagging efficiency used to propagate to the signal region by means of equation (3.11) is obtained as a convolution of the measured efficiency ϵ_b as function of $|\eta|$ with the predicted $|\eta|$ spectrum of the most central jet in simulations; ϵ_b is defined to be zero outside the tracker coverage.

The overall uncertainty on this estimate is determined by the small number of events in the control region and amounts to about 35%, the other systematical uncertainties on ϵ_b and on the extrapolation procedure being negligible in comparison. As the predicted $t\bar{t}$ background after the full selection is only 4.5 events, the impact of this uncertainty is nonetheless subdominant with respect to the statistical uncertainties.

3.7.6. Electroweak WW background

The estimation of the electroweak WW background is performed differently for light and heavy Higgs boson hypotheses, since in the first case it is possible to identify a signal-free region to measure the background directly from the data, while in the second case it is not possible and theoretical predictions have to be used.

Light Higgs boson. For Higgs boson masses up to $200\,\text{GeV}/c^2$, the background is estimated from the observed events in data in the region $m_{\ell\ell} > 100\,\text{GeV}/c^2$, where the signal contamination is negligible. The expected yields from the other background processes in this control region are estimated and subtracted using predictions from data (W + jets, Z + jets, top) or from simulations (W$\gamma^{(*)}$, WZ, ZZ); the total contamination from non-WW processes is 26% in the zero-jet bin, and 54% in the one-jet bin.

The measurement in the control region is then extrapolated to the signal region using the ratio of the predicted WW yields in the signal region and control region from simulations.

$$N_{\text{signal}}^{\text{WW}} = \left(N_{\text{control}} - N_{\text{control}}^{\text{other}}\right) \times \left[\frac{N_{\text{signal}}^{\text{WW}}}{N_{\text{control}}^{\text{WW}}}\right]_{\text{sim}}, \qquad (3.12)$$

To minimize the theoretical uncertainties related to the missing higher order QCD terms in the perturbative series, the estimation is performed separately in each jet multiplicity category. In addition to the statistical and systematical uncertainties from the determination of the WW yield in the control region, a theoretical uncertainty on the extrapolation procedure is also included, estimated as the difference between the extrapolation factors computed with MADGRAPH and MC@NLO(6%).

As the procedure does not allow for a separate determination of the two production modes qq → WW and gg → WW, the theoretical prediction on the ratio between the two cross section is assumed when estimating this background from data. An additional systematical uncertainty is then assigned to the normalization of the gg → WW component of the background, equal to the theoretical uncertainty on the cross section itself (30%). As the gg → WW process amounts to about 5% of the overall WW background yield, this additional uncertainty has a very small impact on the overall sensitivity.

Heavy Higgs boson. For Higgs boson masses above $200\,\text{GeV}/c^2$, the normalization of the WW background is estimated from the theoretical predictions on the production cross section. The theoretical uncertainties for the $qq \rightarrow WW$ cross section in jet multiplicity bins are modelled in terms of the inclusive cross sections for $WW + \geq n$ jets, in the same way as for $gg \rightarrow H$ (Section 3.6.2); uncertainties of 3.4%, 15% and 42% are assumed on $\sigma_{\geq 0}$, $\sigma_{\geq 1}$ and $\sigma_{\geq 2}$ respectively. As in the low mass case, a 30% uncertainty is assumed on the overall normalization of the smaller $gg \rightarrow WW$ background.

Shape uncertainties. For both light and heavy Higgs boson mass hypothesis, the systematical uncertainties on the kinematic of the WW background in the signal region are also included. The uncertainties have been estimated by comparing the output distribution of the multivariate classifier used for signal extraction for events generated with MADGRAPH and with MC@NLO; the former distribution is used for the nominal result. An additional uncertainty is estimated by varying the renormalization and factorization scales in MC@NLO up and down by a factor two simultaneously with respect to the nominal scale. These two uncertainties are taken to be independent, and therefore are used to define two separate possible deformations of the background shape in the signal extraction.

3.7.7. WZ and ZZ background

Background events from diboson production in which both leptons arise from the same Z boson are suppressed by the veto on dilepton invariant masses close to that of an on-shell Z boson, and other leptonic final states from WZ and ZZ are mostly rejected by the requirement that there be only two leptons in the event. As a consequence, the expected contribution from these process to the overall background yield is at the few percent level.

The predictions for diboson backgrounds in the analysis are taken from simulations, using MADGRAPH and PYTHIA for the WZ and ZZ processes respectively. The production cross sections for the two processes are taken from NLO theoretical calculations, with an associated systematical uncertainty of 4% from the higher order terms in the perturbative expansion and of 4% from the knowledge of the parton density functions. As the two processes are closely related from the theoretical point of view, these two uncertainties are assumed to be correlated. As for signal and any other background estimated from simulations, corrections are applied for the reconstruction and trigger efficiencies measured in data and the corresponding systematical uncertainties are are included, as described in Section 3.6.

3.8. Results

2011. An summary of the observed event yields after the full selection for the 2011 dataset and the expectations for signal and background in the different final states, with their associated uncertainties, is given in Table 3.3, and the corresponding distributions of the MVA output discriminator are shown in Figure 3.14. Both figures and table refer to a Higgs boson mass of 130 GeV/c^2, and can serve as prototypical example of the search in all the low mass range. The data is fairly compatible with the background expectations, but a small excess in the signal region at large values of the MVA discriminator is visible.

2012. The observed and expected event yields after the full selection used in the 2012 analysis for a Higgs boson mass hypothesis of 125 GeV/c^2 are shown in Table 3.4. Also in this dataset, a small excess of events is observed with respect to the expectations from the background-only hypothesis. For illustration, the distribution of the transverse mass for the selected events is also shown (Figure 3.15): especially in the more sensitive 0-jet bin, also the shape of the excess is fairly compatible with the expectations from a Higgs boson signal.

	0-jet e µ	0-jet ee+ µ µ	1-jet e µ	1-jet ee+ µ µ	2-jet any
gg → H	40.6 ± 8.9	22.6 ± 4.9	15.9 ± 5.3	6.8 ± 2.3	0.6 ± 0.2
VBF H	0.4 ± 0.1	0.2 ± 0.0	1.7 ± 0.1	0.7 ± 0.1	2.4 ± 0.3
W + H	0.5 ± 0.1	0.3 ± 0.0	0.7 ± 0.1	0.3 ± 0.0	0.0 ± 0.0
Z + H	0.2 ± 0.0	0.1 ± 0.1	0.2 ± 0.0	0.1 ± 0.0	0.0 ± 0.0
qq → WW	209.4 ± 22.6	145.4 ± 15.7	67.5 ± 10.3	34.2 ± 5.2	1.9 ± 0.4
gg → WW	10.5 ± 3.3	6.9 ± 2.2	3.7 ± 1.2	2.1 ± 0.7	0.3 ± 0.1
tt, tW	15.5 ± 3.2	9.8 ± 2.1	43.1 ± 2.6	23.3 ± 1.4	4.5 ± 1.6
W + jets	33.1 ± 11.9	11.6 ± 4.2	20.7 ± 7.4	5.6 ± 2.0	0.6 ± 0.5
Z + jets	0.7 ± 0.1	13.0 ± 6.2	1.0 ± 0.1	28.0 ± 5.2	3.8 ± 4.0
Z → τ+τ−	0.9 ± 0.3	0.0 ± 0.0	2.6 ± 0.6	0.0 ± 0.0	1.0 ± 0.5
W + γ(*)	9.2 ± 2.9	0.9 ± 0.3	1.7 ± 0.6	1.9 ± 1.9	0.2 ± 0.2
WZ, ZZ	4.1 ± 0.4	2.5 ± 0.2	4.8 ± 0.4	1.8 ± 0.1	0.1 ± 0.0
All sig.	41.7 ± 8.9	23.2 ± 4.9	18.5 ± 5.3	7.9 ± 2.3	3.0 ± 0.4
All bkg.	283.4 ± 6.7	190.1 ± 5.6	145.1 ± 4.8	96.9 ± 4.1	12.4 ± 2.7
Data	285	199	145	97	9

Table 3.3. Expected signal and background events for the 2011 analysis after the final selection requirements for a Higgs boson mass hypothesis of 130 GeV/c^2. In addition to the normalization uncertainties reported in this table, uncertainties on the shapes of the MVA classifier distributions are also taken into account.

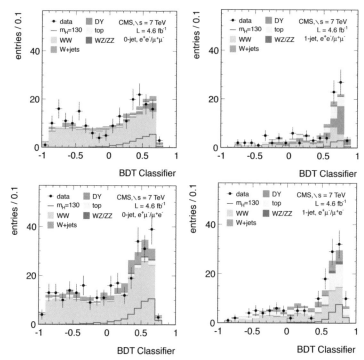

Figure 3.14. BDT classifier distributions for the data and the expected signal and background events for the $m_H = 130 \, \text{GeV}/c^2$ hypothesis: (upper-left) 0-jet bin same-flavour final state, (upper-right) 1-jet bin same-flavour final state, (lower-left) 0-jet bin opposite-flavour final state, (lower-right) 1-jet bin opposite-flavour final state.

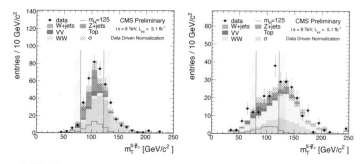

Figure 3.15. Transverse mass distribution for the events in the 2012 data passing the full selection optimized for a Higgs boson mass hypothesis of $125 \, \text{GeV}/c^2$, except for the m_T requirement. The distribution includes both same-flavour and opposite-flavour events, for the 0-jet and 1-jet bins in the left and right panel respectively. The overall uncertainty on the background prediction is shown as a dark grey hatch around the background histogram. The m_T requirements that complete the selection are indicated by the two vertical lines.

	0-jet eμ	0-jet ee+μμ	1-jet eμ	1-jet ee+μμ	2-jet eμ	2-jet ee+μμ
H → WW	23.9 ± 5.2	14.9 ± 5.3	10.3 ± 3.0	4.4 ± 1.3	1.5 ± 0.2	0.8 ± 0.1
WW	87.6 ± 9.5	60.4 ± 6.7	19.5 ± 3.7	9.7 ± 1.9	0.5 ± 0.2	0.3 ± 0.1
$t\bar{t}$, tW	9.3 ± 2.7	1.9 ± 0.5	22.3 ± 2.0	9.5 ± 1.1	3.4 ± 1.9	2.0 ± 1.2
W + jets	19.1 ± 7.2	10.8 ± 4.3	11.7 ± 4.6	3.9 ± 1.7	0.3 ± 0.3	0.0 ± 0.0
Z, WZ, ZZ	2.2 ± 0.2	37.7 ± 12.5	2.4 ± 0.3	8.7 ± 4.9	0.1 ± 0.0	3.1 ± 1.8
W + $\gamma^{(*)}$	6.0 ± 2.3	4.6 ± 2.5	5.9 ± 3.2	1.3 ± 1.2	0.0 ± 0.0	0.0 ± 0.0
All bkg.	124 ± 12	116 ± 15	62 ± 7	33 ± 6	4.1 ± 1.9	5.4 ± 2.2
Data	158	123	54	43	6	7

Table 3.4. Expected signal and background events for the 2012 analysis after the final selection requirements for a Higgs boson mass hypothesis of 125 GeV/c^2.

Chapter 4
Searches for a Higgs boson in the
$H \rightarrow ZZ \rightarrow 4\ell$ decay channels

The $H \rightarrow ZZ \rightarrow 4\ell$ decay mode with electrons and muons in the final state has since ever been considered the golden mode in the searches for a Higgs boson at the LHC: the requirement of four isolated leptons effectively suppresses any background process except for the irreducible continuum electroweak ZZ production, and the excellent resolution on the invariant mass of the system makes it possible to separate the signal from this background. The main limiting factors to the sensitivity in this channel are the small branching fraction of the decay and, for light Higgs bosons, the experimental challenges in reconstructing leptons of very low momentum.

The CMS search in this decay mode is performed inclusively, and re-lies only on the detection of a $\ell^+\ell^-\ell'^+\ell'^-$ final state from a possible ZZ decay. One or both Z bosons are allowed to be off-shell, and no other re-quirements are applied *e.g.* on hadronic jets or missing transverse energy. The other backgrounds besides the irreducible ZZ continuum are Z+jets and $t\bar{t}$, with the leptons arising from the decays of heavy flavour hadrons, photon conversions or misidentified charged hadrons. The signal extrac-tion is performed using a two-dimensional maximum likelihood fit to the invariant mass of the four lepton system and a kinematic discriminator built from the decay angles and the masses of the two $Z^{(*)}$ bosons.

A separate search is performed in the $H \rightarrow ZZ \rightarrow 2\ell 2\tau$ decay channel, as described in the last section of this chapter: a different selection is needed due to the larger reducible backgrounds from jets misidentified as hadronically decaying taus, and to the presence of undetected neutrinos from tau decays in the final state.

4.1. Event selection

For this search events are collected using dielectron and dimuon triggers with asymmetric p_T thresholds of $17/8\,\mathrm{GeV}/c$, and by loose identification and isolation requirements in the case of electrons. The trigger efficiency is above 99% for 4e and 4 μ final states in which the leptons are within the p_T and η acceptance of the analysis; the efficiency for 2e2 μ final states is slightly smaller for light Higgs boson hypotheses, *e.g.* 97.5% for $m_H = 120$ GeV, but eventually reaches a plateau value above 99% for $m_H > 140$ GeV. In the 2012 analysis, these two triggers were complemented by the electron-plus-muon trigger with p_T thresholds of 17 GeV/c on the leading lepton and 8 GeV/c on the subleading one, to recover this small inefficiency in the 2e2 μ final state.

4.1.1. Lepton selection

Among all the SM Higgs boson searches, the H → ZZ → 4ℓ analysis is the most demanding in terms of lepton reconstruction efficiency, especially at low momentum since for light Higgs boson masses the p_T spectrum of the trailing lepton is very soft (Figure 4.1).

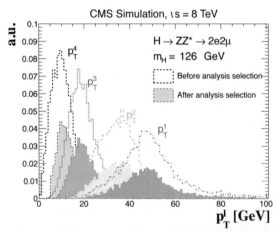

Figure 4.1. Distribution of the transverse momentum for each of the four leptons, ordered in p_T, from H → ZZ → 2e2 μ signal events for a Higgs boson mass hypothesis 126 GeV/c^2. The distribution is shown both for all simulated leptons within the η acceptance (empty histograms), and for reconstructed and selected events (shaded histograms).

The offline selection requires the presence of two pairs of electrons or muons, within the acceptance requirements $|\eta| < 2.5$ and $p_T > 7$ GeV/c for electrons, and $|\eta| < 2.4$ and $p_T > 5$ GeV/c for muons. To preserve a good efficiency down to low transverse momentum, the loose muon identification requirements described in Section 2.4.3 are used, corresponding to a per-lepton reconstruction and identification efficiency above 95% for $p_T > 5$ GeV/c, and reaching a plateau value above 99% for $p_T > 10$ GeV/c.

For electrons, a multivariate identification algorithm is used, based on a combination of calorimetric and tracking information, optimized separately in three categories of electron pseudorapidity and two ranges in transverse momentum. To further suppress electrons from photon conversions, electron tracks missing more than one hit in the innermost tracker layers are rejected unless the extrapolation of the track is not compatible with any active sensor in those layers. The combined reconstruction and identification efficiency for electrons is about 65% for $p_T < 10\,\text{GeV}/c$, and reaches a plateau of about 90% for $p_T \sim 20\,\text{GeV}/c$. In this analysis electrons in the transition region between barrel and endcaps are also included; the overall efficiency for electrons in this region is about 5% less than for those in the fiducial acceptance of the ECAL.

Impact parameter. In order to reject non-primary electrons and muons, a requirement on the significance of the 3D impact parameter is applied, $S_{IP} < 4$. This requirement is relaxed to $S_{IP} < 100$ in some of the control regions used to study the reducible backgrounds, loose enough to retain leptons from the decays of heavy flavour hadrons but still rejecting most of the activity from pile-up interactions. To further reject tracks with poorly measured impact parameter, *e.g.* electrons from photon conversions or muons from in-flight decays of light hadrons, absolute requirements on the longitudinal and transverse impact parameter are also applied, $d_z < 1\,\text{cm}$, $d_{xz} < 0.5\,\text{cm}$; as for the $S_{IP} < 100$ criteria, the requirements are loose enough to preserve efficiency also for leptons from the decays of heavy flavour hadrons in the control regions.

Lepton isolation. Lepton isolation criteria are defined using the candidates reconstructed by the particle flow algorithm, clustered in cones of radius $R = \sqrt{\Delta\eta^2 + \Delta\phi^2} = 0.4$ around the leptons. Particle candidates that are identified as electrons or muons by the particle flow selection are excluded from the isolation cone. Photon candidates identified in the context of final state radiation recovery, explained later, are also excluded from the isolation sum.

A combined relative isolation variable is build from the sum of the p_T of the charged hadrons, neutral hadrons and photons in the isolation cone, divided by the transverse momentum of the lepton. The contamination from pile-up in the neutral component of the isolation is subtracted statistically using the effective area method, as described in Section 2.4.4, leading to a stable efficiency as function of pile-up multiplicity (Figure 4.2).

Lepton disambiguation. Due to the loose selection criteria used, especially when the identification requirements are relaxed to measure the

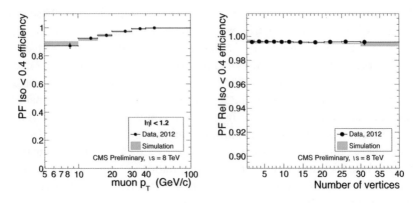

Figure 4.2. Tag-and-probe results for muon isolation efficiencies in 2012 data and simulations, measured from $Z \rightarrow \mu^+ \mu^-$ events: efficiency as function of p_T in the barrel region (left), and average efficiency as function of the multiplicity of primary vertices for $p_T > 20\,\mathrm{GeV}/c$ (right).

reducible backgrounds, it is possible for the same particle to be reconstructed both as an electron candidate and as a muon candidate; typically this happens if an energy deposit is found in the ECAL in the proximity of a muon, due to final state radiation, interactions in the material or contamination from another particle in the event. This ambiguity is resolved in the analysis by vetoing electrons which are within a cone of radius $R = \sqrt{\Delta\eta^2 + \Delta\phi^2} = 0.05$ from a muon that passes the Particle Flow identification criteria or is reconstructed as a global muon (Section 2.4.3).

In addition, it can happen that multiple reconstructed muon candidates are created from a single signal in the muon system, due to two possible effects: (i) tracks from other charged hardons in the event that happen to be loosely compatible with the reconstructed segment from a muon in the event; (ii) cases where a muon track in the inner detector is reconstructed multiple times from the different tracking iterations and, due to inefficiencies in the pattern recognition, the fraction of overlapping hits between the two tracks is too small for them to be correctly identified as ambiguous[1]. The first category of duplicates is removed from the analysis by assigning the muon segments only to the track that is most

---

[1] In the majority of these cases, one track is reconstructed early in the sequence, and contains hits in the pixel detector and up to a few hits in the inner silicon strip layers due to some failure in further extending the trajectory building. The track is then found again at later steps of the reconstruction seeded from the outer tracker layers, but cannot be extended up to the center of the detector since the hits in the inner layers have already been masked as used by the first track. The probability for such failure to happen, measured for muons from $Z \rightarrow \mu^+ \mu^-$ decays in data, is about 0.5%.

compatible with them, while the second category is dealt with in the combinatorial reconstruction of the events by requiring all muons to have an angular separation $\Delta R > 0.02$ in the (η, ϕ) space. While the fraction of good muons which result in duplicates is at the few per mill level, this disambiguation is very important in the control regions with multiple leptons with relaxed requirements, as the suppression in the production cross section for SM processes with additional good leptons in the final state is of comparable order: e.g. before any disambiguation criteria, an inclusive sample of tri-muon events from $(Z \rightarrow \mu^+ \mu^-) + \mu$ candidates with relaxed selection contains a contamination of about 15% and 50% from muon duplicates of kind (i) and (ii) respectively, which is reduced to about 1% by the disambiguation criteria.

4.1.2. Final state radiation recovery

The probability for a photon to be emitted in a $Z \rightarrow \ell^+\ell^-$ decay, when a $p_T > 2\,\text{GeV}$ threshold is applied to the photon transverse momentum, is about 8% for dimuon decays and 15% for dielectron decays. In the case of the electrons usually the energy of the photon is already included in the ECAL energy measurement as the photon emission is mostly collinear with the lepton; however, this does not happen for muons, or for photon emission at larger angles.

The identification of photons from final state radiation (FSR) is important in the analysis in two respects: (i) the inclusion of FSR photons in the computation of the invariant mass of the system results in a sharper peak for the Higgs boson signal compared to what achievable using only the momenta of the four leptons; (ii) the removal of FSR photons from the isolation cone of the leptons improves the selection efficiency. As the rate of photons is small, it is of paramount important to achieve a high purity in the identification, since the erroneous association of a spurious photon, e.g. from a π^0 decay, to a non-radiative signal event will result in a mismeasurement of the mass, thereby degrading the sensitivity.

Photon selection. The soft p_T spectrum of FSR photons makes the task particularly challenging from the experimental point of view. The traditional photon reconstruction algorithms used e.g. in the $H \rightarrow \gamma\gamma$ search are not applicable in this context, as they are aimed at high energy isolated objects. The low energy photon reconstruction in the particle flow algorithm is used instead, allowing the reconstruction of photons down to a p_T of $2\,\text{GeV}/c$ with an efficiency of about 50% (Figure 4.3, left).

Energy deposits in the ECAL associated to a muon track are also promoted to photon candidates, if their transverse energy exceeds $2\,\text{GeV}$;

these extra photons contribute by about 20% to the overall rate of reconstructed FSR photons in simulated H → 4 μ signal events.

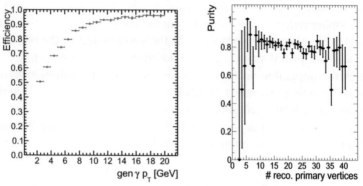

Figure 4.3. Reconstruction efficiency (left) and purity (right) for FSR photons
in simulated H → 4 μ events.

To avoid double counting with the electron supercluster and contaminations from secondaries produced in the interactions in the detector material, photon candidates are vetoed in the proximity of an electron, if
$\Delta R < 0.15$ or if $\Delta \eta < 0.05$ and $\Delta \phi < 2.0$. The photon selection
criteria are optimized as function of the angular distance between the
photon and the closest lepton in (η, ϕ) space. In an inner cone $R \leq 0.07$,
photons are accepted if $p_T > 2 \, \text{GeV}/c$, with no further requirements. In
an outer annulus $0.07 < R < 0.5$, where the rate of photons from underlying event and pile-up is much larger, a tighter threshold of $4 \, \text{GeV}/c$ is
used, and the photons are also required to be isolated: the sum of the p_T
of all charged hadrons, neutral hadrons and photons in a cone of radius
$R = 0.3$ centered on the photon should not exceed the p_T of the photon
itself; contrarily to the lepton case, for photon isolation also the charged
hadrons associated to other primary vertices are used, since the photon
might come from a pile-up event, and no subtraction is applied to the
neutral part. The selection criteria have been tuned to achieve approximately the same purity in the two angular regions.

Radiative decay reconstruction. Photon candidates that satisfy the
FSR selection described above are provisionally associated to the closest
lepton. Subsequently, for each of the two dilepton pairs in a candidate
H → ZZ → 4ℓ decay, all candidate FSR photons assigned to either
lepton are considered individually, and retained only if they satisfy the
two additional requirements $m_{\ell\ell\gamma} < 100 \, \text{GeV}$ and $|m_{\ell\ell\gamma} - m_Z| < |m_{\ell\ell} - m_Z|$, where m_Z denotes the nominal Z boson mass. If multiple photon
candidates satisfy also these requirements, only the best candidate is se-

lected; the best candidate is defined as the one with highest p_T, unless all photon candidates have $p_T < 4\,\text{GeV}$; in the latter case, the best photon is defined as the one with with minimal ΔR with respect to the closest lepton.

After the full selection including the $m_{\ell\ell\gamma}$ requirements, the purity of the selection amounts to about 80% (Figure 4.3, right). The purity is stable with respect to the pile-up multiplicity, since the photon isolation criteria become progressively tighter because of the pile-up contributions in the isolation cone. The procedure was validated on $Z \rightarrow \ell^+\ell^-$ events in 2011 data, where the observed rate of identified radiative decays was found to be in good agreement with the predictions from simulation.

4.1.3. Combinatorial reconstruction

The first step of the combinatorial reconstruction is the search for $\ell^+\ell^-$ pairs of same-flavour opposite-charge leptons. For each of such pairs, made from leptons satisfying the identification and impact parameter requirements, the final state radiation recovery procedure described in the previous section is applied, and the pair is retained if both leptons satisfy the isolation criteria.

Among all possible lepton pairs, the one with dilepton invariant mass closest to the nominal Z boson mass is selected, and later referred to as Z_1. A requirement is applied on the invariant mass of this pair, $40 < m_{Z_1} < 120\,\text{GeV}/c^2$, where in case of radiative decays the photon four-momentum is included in the computation of the invariant mass.

A second, non overlapping, dilepton pair Z_2, made of different leptons, is then selected to build a four-lepton candidate. If more than one dilepton pair candidate is found, the Z_2 is selected as the one with the largest scalar sum of the p_T of the two leptons. To achieve optimal sensitivity for a light Higgs boson, a looser requirement is applied to the invariant mass of the second pair, $12 < m_{Z_2} < 120\,\text{GeV}/c^2$.

To ensure that the events are on the plateau of the trigger efficiency, it is required that among the four leptons in the ZZ candidate at least one has $p_T > 20\,\text{GeV}/c$ and at least another one has $p_T > 10\,\text{GeV}/c$.

In order to further suppress those reducible background events where a dilepton pair arises from the decay of a heavy flavour hadron, an additional requirement $m_{\ell\ell'} > 4\,\text{GeV}/c^2$ is applied to all four opposite-charge pairs of leptons in the ZZ candidate, irrespectively of their flavour and their assignment to the Z_1 or Z_2.

The search region for a SM Higgs boson is eventually defined as a requirement on the m_{ZZ} invariant mass, dependent on the Higgs boson mass hypothesis considered.

The observed and expected event yields at various steps of the selection are shown in Figure 4.4 for the 7 and 8 TeV datasets in the 2e 2 μ final state. In these comparisons, the expected background yields are estimated from simulations for all processes; the simulated events are re-weighted to match the pile-up distribution observed in data, but no other correction is applied *e.g.* for selection efficiencies. At the earliest step of the analysis, the observed yield in data is larger than the expectations since the latter does not include the contributions from QCD multi-jet production: this contamination is drastically reduced by the requirement of a Z candidate from leptons satisfying identification and isolation requirements, resulting in a very good agreement between observation and expectations.

A small discrepancy can also be seen in an intermediate step requiring a Z_1 candidate plus a third lepton. A disagreement at this level is not unexpected since the yield at this step is dominated by Z + jets background events, and is sensitive to the production cross section of Z plus heavy flavour hadrons and to the modelling of the jet-induced leptons, none of which is expected to be accurate to better than $O(10\%)$. The very good agreement between observation and expectations is once again restored at the subsequent step of the selection, when the yield is eventually driven by the electroweak ZZ production.

A summary of the expected and observed event yields for the combination of the 7 and 8 TeV datasets is provided in Tables 4.1 and 4.2 for the low mass range and full mass range respectively. In this case, the expected yields for the reducible background are estimated from data.

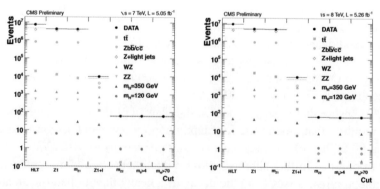

Figure 4.4. Expected and observed event yields for the various steps of the H → ZZ → 4ℓ selection in the 2e 2 μ final state, for the 2011 data (left) and 2012 data (right). The $Z_1 + \ell$ step refers to the requirement of a third lepton passing the nominal selection. In these figures, predictions from simulations are used for all processes, including the Z+jets backgrounds. In the three rightmost bins of both plots, the marker for the ZZ yield is not visible as it is almost exactly behind the point corresponding to the observed yield in data.

	4e	4 μ	2e 2 μ
$m_H = 120\,\text{GeV}/c^2$	0.8 (18%)	1.6 (36%)	1.9 (24%)
$m_H = 126\,\text{GeV}/c^2$	1.5 (21%)	3.0 (42%)	3.8 (39%)
$m_H = 130\,\text{GeV}/c^2$	2.1 (23%)	4.1 (45%)	5.4 (32%)
ZZ continuum	2.7 ± 0.3	5.7 ± 0.6	7.2 ± 0.8
reducible bkg.	$1.2^{+1.1}_{-0.8}$	$0.9^{+0.7}_{-0.6}$	$2.3^{+1.8}_{-1.4}$
observed events	6	6	9

Table 4.1. Expected and observed event yields after the H $\to$ ZZ $\to$ 4ℓ selection in the low mass range $110 < m_{4ZZ} < 160\,\text{GeV}/c^2$, for SM Higgs boson signals of different masses, the irreducible ZZ continuum and the reducible backgrounds (dominated by $Z + X$). The overall acceptance times detection efficiency for signal events is also given in parenthesis next to the signal yields. The yields correspond to integrated luminosities of $5.05\,\text{fb}^{-1}$ at $\sqrt{s} = 7\,\text{TeV}$ and $5.26\,\text{fb}^{-1}$ at $\sqrt{s} = 8\,\text{TeV}$. The reducible background prediction is derived from data, as described in the text.

	4e	4 μ	2e 2 μ
$m_H = 200\,\text{GeV}/c^2$	8.3 (39%)	13.3 (62%)	21.6 (51%)
$m_H = 350\,\text{GeV}/c^2$	4.8 (42%)	7.5 (66%)	12.7 (58%)
$m_H = 500\,\text{GeV}/c^2$	1.7 (46%)	2.6 (71%)	4.4 (60%)
ZZ continuum	29.3 ± 3.4	49.0 ± 5.1	75.5 ± 8.0
reducible bkg.	$3.0^{+2.7}_{-1.9}$	$2.2^{+1.6}_{-1.3}$	$5.0^{+4.0}_{-3.0}$
observed events	32	47	93

Table 4.2. Expected and observed event yields after the H $\to$ ZZ $\to$ 4ℓ selection in the full mass range $100 < m_{4ZZ} < 800\,\text{GeV}/c^2$, for SM Higgs boson signals of different masses, the irreducible ZZ continuum and the reducible backgrounds (dominated by $Z + X$). The overall acceptance times detection efficiency for signal events is also given in parenthesis next to the signal yields. The yields correspond to integrated luminosities of $5.05\,\text{fb}^{-1}$ at $\sqrt{s} = 7\,\text{TeV}$ and $5.26\,\text{fb}^{-1}$ at $\sqrt{s} = 8\,\text{TeV}$. The reducible background prediction is derived from data, as described in the text.

4.2. Matrix element likelihood analysis

While the invariant mass distribution of the ZZ candidates is the most natural discriminating variable to resolve a H $\to$ ZZ $\to$ 4ℓ signal over the continuum ZZ background, the sensitivity of the search can be increased by exploiting also the different angular distribution of the four leptons [87, 88].

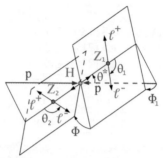

Figure 4.5. Definition of the five angles θ^*, Φ_1, θ_1, θ_2, Φ for the pp $\rightarrow$ H $\rightarrow$ ZZ $\rightarrow$ $4\ell4$ process. The two angles two production angles θ^* and Φ_1 are defined in the H rest frame, while the angles θ_1, θ_2, and Φ are defined in the Z_i rest frames. [87]

A suitable parametrization of the decay kinematic can be given in terms of five angles θ^*, Φ_1, θ_1, θ_2, Φ (Figure 4.5). Below the threshold for the production of two on-shell Z bosons, the invariant mass of the two dilepton pairs m_{Z_1} and m_{Z_2} provides further discriminating power.

All these variables can be combined into a kinematic discriminator $K_D = P_{sig}/(P_{sig} + P_{bkg})$, where P_{sig} and P_{bkg} are the conditional signal and background probability density functions in those seven discriminating variables at fixed value of m_{ZZ}. For P_{sig} the phase-space and Z propagator terms are included in a fully analytic parameterization [89, 87], while P_{bkg} is derived empirically from simulated q$\bar{q}$ $\rightarrow$ ZZ events. The distributions of the kinematic discriminator and of three of the most sensitive individual variables is shown in Figure 4.6.

The analytical computation for the decay kinematic used to build the likelihood discriminator is performed at leading order, so photon emission from final state is not included. To cope with this limitation, when evaluating the likelihood discriminator in events where a final state radiation photon is identified, the four-momentum of the photon is added to the four-momentum of the closest lepton. This is a good approximation to the extent that final state radiation is mostly collinear or soft.

The H $\rightarrow$ ZZ $\rightarrow$ 4ℓ search is implemented as a bidimensional unbinned analysis, where each individual event is characterized by its invariant mass m_{ZZ} and the value of the kinematic discriminator K_D. The bidimensional probability density function for each process is defined as a conditional product

$$p(m_{ZZ}, K_D) = p_m(m_{ZZ}) \cdot p_{K_D}(K_D \mid m_{ZZ}) , \qquad (4.1)$$

where $p_m(m_{ZZ})$ is an analytical function describing the expected mass spectrum from the given process, and $p_{K_D}(K_D \mid m_{ZZ})$ is a conditional

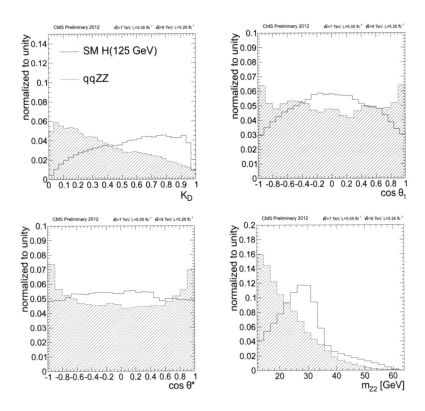

Figure 4.6. Distribution of the kinematic discriminator $K_D = P_{sig}/(P_{sig} + P_{bkg})$ and three of the most sensitive variables used in it: $\cos\theta_1$, $\cos\theta^*$, m_{Z_2}.

pdf for K_D at each value of m_{ZZ}. In order to account for detector effects, and to benefit from the most accurate description available in the NLO event generators, the $p_{K_D}(K_D \mid m_{ZZ})$ is obtained from simulated events, in the form of a binned bidimensional template.

In the case of signal events, the angular distribution obtained from the POWHEG event generator is compared to the more accurate predictions from PROFECY4F, which includes interference and NLO corrections to the decay, and found to be compatible.

4.3. Signal lineshape modelling

For light Higgs bosons, the invariant mass distribution of the signal is determined only by the experimental resolution, since the nominal width of the Higgs boson is exceedingly narrow. Therefore, in the analysis for $m_H < 160\,\text{GeV}/c^2$, the signal mass distribution is modelled with a Crystal Ball function whose parameters are determined from a fit to the simulated signal events separately for several Higgs boson mass hypotheses, and then smoothly interpolated to all other m_H values.

For heavier Higgs boson hypotheses, the signal distribution is modelled with the convolution of parametric description of the expected Higgs boson lineshape and a Crystall Ball function accounting for the detector response. The parametrization for the theoretical lineshape $d\sigma/dm_{ZZ}$ is a Breit-Wigner distribution with running Higgs boson width, multiplied with the gluon-gluon partonic luminosity $dL_{gg}(\hat{s})/d\hat{s}$,

$$
f(m_{ZZ}|m_H) = \frac{1}{\mathcal{N}} \frac{m_{ZZ}\,\Gamma_{gg}(m_{ZZ})\,\Gamma_{ZZ}(m_{ZZ})}{(m_{ZZ}^2 - m_H^2)^2 + m_{ZZ}^2\,\Gamma_{tot}(m_{ZZ})} \times \left[\frac{dL_{gg}}{d\hat{s}}\right]_{\hat{s}=m_{ZZ}^2}, \quad (4.2)
$$

where m_H is the nominal mass of the Higgs boson.

Systematical uncertainties on the electron and muon energy scales and resolutions are implemented as uncertainties on the parameters of the response model. Residual corrections due to the different resolutions in data and simulation are also applied, determined from the fits to the Z $\rightarrow$ $\ell^+\ell^-$ lineshape.

Event-by-event mass resolution.   The lineshape model used in the analysis represents the average mass resolution over an ensemble of H $\rightarrow$ ZZ $\rightarrow$ 4ℓ events. However the spread of mass resolution across the individual events is large, varying from about half to about twice the typical resolution dependent on the lepton p_T and η, so for the interpretation of the results it is also interesting to have an estimate of the mass resolution in the observed events in data.

An estimate on the mass resolution for an events can be obtained by propagating the estimated resolutions on the four-momenta of the individual leptons and FSR photons. For a final state with n particles and using cartesian coordinates, a $(4\,n) \times (4\,n)$ covariance matrix C of the individual momentum components can be written in block-diagonal form starting from the 4×4 covariance matrices C_i of the components of each four-momentum; the Jacobian matrix J to project from the full covariance matrix to the resolution on the invariant mass is easily derived:

$$
J = \left(\frac{\partial m^2}{\partial E_1}, \frac{\partial m^2}{\partial p_{x,1}}, \dots, \frac{\partial m^2}{\partial p_{z,1}}; \dots \dots; \frac{\partial m^2}{\partial E_n}, \frac{\partial m^2}{\partial p_{x,n}}, \dots, \frac{\partial m^2}{\partial p_{z,n}}\right), \quad (4.3)
$$

whose components are

$$
\frac{\partial m^2}{\partial E_i} = \frac{\partial}{\partial E_i}\left(\left(\sum_i E_i\right)^2 - \sum_a\left(\sum_i p_{a,i}\right)^2\right) = 2\sum_i E_i
$$

$$
\frac{\partial m^2}{\partial p_{a,i}} = \frac{\partial}{\partial E_i}\left(\left(\sum_i E_i\right)^2 - \sum_b\left(\sum_i p_{b,i}\right)^2\right) = -2\sum_i p_{a,i},
$$

where the index i runs over the particles, and the subscripts a, b run over the components of the three-momentum. The estimate of the mass resolution is obtained from a similarity transformation from J and C:

$$\delta m = \frac{1}{2m} \sqrt{J^t \, C \, J} \, . \tag{4.4}$$

For muons, the covariance matrix of the components of the three-momentum are taken from the Kalman Filter fit of the track, and extended to a 4×4 matrix, of rank 3, relying on the relation $\partial E / \partial p_a = p_a / E$. For electrons and photons, the resolution on the angles are neglected and the uncertainty is estimated from a parametrization dependent on the p_T and η of the particle; a 4×4 matrix, of rank one, is obtained similarly by $\partial p_a / \partial E = E / p_a$.

Studies on simulated events show that the estimate of the mass resolution from equation (4.4) underpredicts the actual resolution; this is partially due to the non-perfect tuning of the per-lepton uncertainties and partially due to the non-Gaussian tails in the resolution on the individual components, which are to some extent reabsorbed in the Gaussian core of the combined mass resolution as expected from the central limit theorem. This discrepancy is corrected by rescaling the estimated mass resolution by a scale factor dependent on the invariant mass of the system; at low mass, the correction factor is about 1.3.

4.4. Background estimation and modelling

The major background in this search is the electroweak continuum ZZ production (Figure 4.7). Reducible backgrounds due to lepton misidentification, mostly from Z + jets, are nonetheless relevant especially in the low mass region.

Figure 4.7. Feynman diagrams for the irreducible ZZ background: the dominant $q\bar{q} \to ZZ$ process (left), and smaller gluon-induced contribution $gg \to ZZ$.

4.4.1. ZZ background

The irreducible ZZ continuum is estimated from simulated events, normalized to the NLO calculations from MCFM [90, 91]. For the dominant production mode $q\bar{q} \to ZZ$ events are generated at NLO accuracy with POWHEG [57, 92], while for the $gg \to ZZ$ production the leading order GG2ZZ [93] generator is used. The $gg \to ZZ$ production modes is formally a next-to-next-to-leading order compared to $q\bar{q} \to ZZ$, but it amounts to about 8.5% of the overall ZZ cross section at $\sqrt{s} = 7\,\text{TeV}$ due to the larger gluon partonic luminosity at LHC; the contribution is slightly larger at 8 TeV, 9.5%, due to the different scaling of the two partonic luminosities as function of the energy.

The expected m_{ZZ} distributions for $q\bar{q} \to ZZ$ and $gg \to ZZ$ extracted from simulated events are then parametrized with a smooth empirical distribution. Since the background distribution does not present sharp features on a mass scale comparable with the experimental width of Higgs boson signal, only uncertainties on the background normalization are considered.

Theoretical uncertainties on the differential cross section $d\sigma/dm_{ZZ}$ from the knowledge of the parton density functions and the neglected higher orders in the perturbative series have been evaluated using MCFM. As the signal extraction for each Higgs boson mass hypothesis is performed considering only a restricted four-lepton mass range, and the dependence of the uncertainty on m_{ZZ} within that range is negligible, the theoretical uncertainty on $d\sigma/dm_{ZZ}$ at the center of the range is applied to the overall normalization of the ZZ background.

The uncertainties from higher orders increase with m_{ZZ}, and are in the 2–6% range for $q\bar{q} \to ZZ$ and in the 20–45% for $gg \to ZZ$. Uncertainties from parton density functions are similarly increasing from 3%/10% at $m_{ZZ} = 120\,\text{GeV}/c^2$ to 8%/16% at $m_H = 600\,\text{GeV}/c^2$, the larger uncertainty figure being again for the $gg \to ZZ$ production mode which relies on the more poorly known gluon partonic luminosities. The size of the uncertainties from both sources is very similar for $\sqrt{s} = 7\,\text{TeV}$ and 8 TeV.

The theoretical uncertainties on the background prediction are assumed to be totally correlated between the 2011 and 2012 dataset, since they rely on the same computation and, in the case of parton density functions, on the same set of input data.

In the past, an alternative estimation based on the ratio of the inclusive Z and ZZ production cross sections has been used, to benefit from the cancellation of the luminosity scale uncertainty and a partial cancellation of the uncertainties on selection efficiencies and parton density functions. It is no longer used in the present analysis, as more precise calibrations of the luminosity scale have become available, competitive with the theoretical uncertainty on the Z production cross section.

While the smoothness of the continuum background could in principle be used to estimate the ZZ yield directly from the sidebands in the reconstructed mass, in practice with the current integrated luminosities this method is not competitive since the statistical uncertainty on the number of events in the sidebands is much larger than the theoretical uncertainty on the background prediction; the overall uncertainty on estimate from simulation is at the 10% level, while the yield for the continuum background is about a dozen events for the full low mass range, 110–160 GeV/c^2 (Table 4.1).

4.4.2. Reducible backgrounds

The overall normalization of all reducible backgrounds is obtained using the technique of misidentification probabilities, similarly to what is done in the H → WW analysis for the W + jets background (Section 3.7.1). Two estimates are obtained with partially independent methods, one based on control regions with same sign leptons, and one on control regions with leptons failing the nominal selection. The two estimates are then combined conservatively by taking the average of the two results as central value and the envelope of the two $\pm 1\sigma$ intervals as uncertainty band.

Loose lepton definition.   Both methods rely on a common definition of loose leptons, obtained from the nominal selection by removing the isolation requirements and relaxing the identification requirements to the most inclusive definition allowed by the muon and electron reconstruction. The requirement on the significance of the impact parameter and the conversion rejection criteria for electrons are preserved, to maintain a flavour composition of the background leptons more similar to that of the signal region. The disambiguation criteria described in Section 4.1.1 are also preserved, to avoid cases where the detector signal from a good lepton gives rise also to additional loose leptons.

Misidentification probabilities.   For both methods, the misidentification probabilities are measured from a three lepton control sample of $Z_1 + \ell$, where only loose requirements are applied on the third lepton. Impact parameter and conversion rejection requirements are preserved, to maintain a flavour composition more similar to the one in the control region. The contamination from WZ and ZZ are suppressed requiring $E_T^{\mathrm{miss}} < 25\,\mathrm{GeV}$, and by vetoing the presence of a fourth lepton.

Similarly to what done in the signal region, a dilepton invariant mass requirement $m_{\ell\ell} > 4\,\mathrm{GeV}/c^2$ is applied to the opposite-charge pair of leptons obtained by combining the third lepton with one of the two leptons of the Z_1. Inclusive misidentification probabilities are obtained from this control sample as the fraction of events where the third lepton satisfies also the nominal selection requirements.

The measured misidentification probability for muons is approximately 10%, with very little dependency on η and p_T, and is dominated from the isolation criteria since the majority of the reconstructed muon candidates even with just the relaxed identification criteria are from real muons, *e.g.* from heavy flavour decays. For electrons, instead, the misidentification probability at low p_T is suppressed by the slower turn-on of the identification efficiency: in the barrel, the probability is about 2% for electrons in the 7–10 GeV/c p_T bin, and reaches a plateau of 6% at $p_T > 15$ GeV/c; the misidentification probability in the endcaps has a similar p_T dependency, but is larger by a factor two due to the worse discrimination in the identification part of the selection.

Same-sign method.   In this approach, a $Z_1 + \ell^{\pm}\ell^{\pm}$ control region is defined by requiring a Z_1 candidate and two additional loose leptons with the same flavour and the same charge, to suppress the ZZ contamination. The same kinematic requirements of the signal region are applied also in the control region: $40 < m_{Z_1} < 120$ GeV/c^2, $12 < m_{Z_2} < 120$ GeV/c^2, and $m_{\ell\ell} > 4$ GeV/c^2 for any other pair of opposite-charge leptons. For this relaxed selection, a fair agreement between data and predictions from simulation is observed, within the limited statistical accuracy, both for the same-charge leptons defining the control region, and also for the opposite-charge ones (Figure 4.8).

The prediction for the reducible background in the signal region is obtained by weighting the events in the control region by the product of the misidentification probabilities for the two leptons $\epsilon_{\text{id+iso}}$, and applying an overall correction factor for the extrapolation from same-sign to opposite-sign determined from simulations, as in eq. (4.5). The correction factor is about 1.3 for the 4 μ final state, motivated by the charge correlation of leptons in $Z + b\bar{b}$ decays, while it is close to unity for the 4e since electrons from jets arise mostly from misidentified hadrons or converted photons from π^0 decays, with no charge correlation; as the reducible background yield in the control region for 2e 2 μ is dominated by events where the two electrons arise from misidentifications, the correction factor is similar to that of 4e.

$$N_{\text{id+iso}}^{Z+\ell^{\pm}\ell^{\mp}} = \epsilon_{\text{id+iso}}^2 \times \left[N_{\text{control}}^{Z+\ell^{\pm}\ell^{\pm}} \right]_{\text{data}} \times \left[\frac{N_{\text{control}}^{Z+\ell^{\pm}\ell^{\mp}}}{N_{\text{control}}^{Z+\ell^{\pm}\ell^{\pm}}} \right]_{\text{sim}} . \qquad (4.5)$$

This background estimate is based on the assumption that the background contains two independent misidentified leptons, and therefore underestimates the contributions from events with three good leptons from WZ + jet, jet $\rightarrow \ell$; however, from studies on simulations this background is found to be only a small fraction of the reducible backgrounds (5–10%).

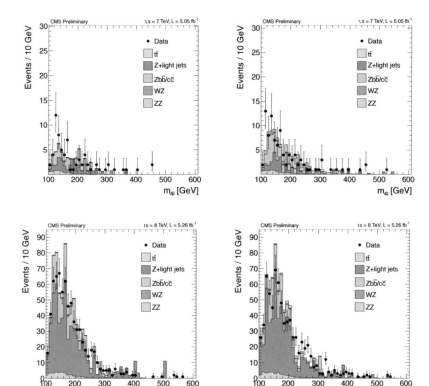

Figure 4.8. Selected comparison between the observed $m_{4\ell}$ distribution in data and the predictions from simulations in the inclusive $Z_1 + \ell\ell$ sample, when the two additional leptons are required to have the same charge (left) or opposite charge (right), and no identification and isolation requirements are applied on those two leptons: $4\,\mu$ events from 2011 data are shown in the top row, and 4e events from 2012 data in the bottom row. In these plots, the prediction for all background distributions and normalizations are taken directly from simulations. The simulation prediction for Z+jets is split into heavy flavour jets (green, labelled as $Z + b\bar{b}/c\bar{c}$) and light flavours (blue, labelled as $Z + jets$) according to the generator level information.

Similarly, correlated e^+e^- pairs from photon conversions in $Z + \gamma$ are not considered, although events in which one of the two electrons arises from conversion are; however, this background is estimated to be negligible from studies on simulations. To be conservative, an additional 10% uncertainty on the reducible background is nonetheless included to account for possible contributions from these two sources.

The statistical uncertainties on the estimated background normalization are 8% for the four-muon final state and 3% for the others, the difference arising from the smaller background yield for muon candidates even before isolation and identification. The systematical uncertainty from the knowledge of the misidentification probabilities is about 30% for the 4e and 2e 2 μ final states, and 40% for the four-muon one.

As the kinematic distribution of same-charge and opposite-charge events is different, the prediction for the shape of the $m_{4\ell}$ spectrum for the reducible backgrounds is determined by applying the misidentification probabilities method to opposite-charge simulated events, and parameterizing the corresponding distribution with a smooth empirical function. As the background distribution does not have sharp features on mass scales comparable to the resolution on the reconstructed four-lepton mass, "horizontal" uncertainties on the shape of the distribution are negligible with respect to the "vertical" normalization uncertainties.

Pass-plus-fail method. A second approach, more similar to what done for W + jets in the H $\rightarrow$ WW $\rightarrow$ $2\ell 2\nu$ analysis, relies on two $Z_1 + \ell^\pm \ell^\mp$ control regions where one or both of the additional leptons fail the nominal isolation and identification requirements, denoted as 2P2F (2 pass, 2 fail) and 3P1F (3 pass, 1 fail).

An alternative set of misidentification probabilities is used in this estimate, obtained from a three lepton control sample with tighter selection criteria on the Z_1 mass, $|m_{Z_1} - m_Z| < 10\,\mathrm{GeV}/c^2$, and on the angular separation between the third lepton and any of the two leptons from the Z_1, $\Delta R > 0.8$, to ensure non-overlapping isolation cones. The misidentification probabilities obtained with this definition are smaller than the inclusive ones, in particular in the case of electrons where the tight mass criteria suppresses radiative Z $\rightarrow$ eeγ decays where the photon is misidentified as a loose lepton, but this is taken into account using the 3P1F control sample.

First, a propagation of the reducible background from the 2P2F sample to the 3P1F is performed by weighting events with the misidentification probabilities for the two leptons,

$$N_{3P1F}^{Z+\mathrm{jets}} = \left[\frac{\epsilon_{\mathrm{id+iso}}(\ell_1)}{1 - \epsilon_{\mathrm{id+iso}}(\ell_1)} + \frac{\epsilon_{\mathrm{id+iso}}(\ell_2)}{1 - \epsilon_{\mathrm{id+iso}}(\ell_2)} \right] \times N_{2P2F}^{\mathrm{data}}. \qquad (4.6)$$

Then, the other reducible 3P1F background not predicted from the 2P2F extrapolation is determined by subtracting from the 3P1F yield in data the prediction from eq. (4.6), and the ZZ contamination obtained from simulations.

$$N_{3P1F}^{\mathrm{other}} = N_{3P1F}^{\mathrm{data}} - \left[N_{3P1F}^{Z+\mathrm{jets}} \right]_{2P2F} - \left[N_{3P1F}^{ZZ} \right]_{\mathrm{sim.}}. \qquad (4.7)$$

Finally, a combined prediction for the reducible background in the signal region is obtained from a combined extrapolation from the 3P1F and 2P2F regions:

$$N_{4P}^{bkg.} = \left[\frac{\epsilon(\ell)}{1 - \epsilon(\ell)} \right] \times N_{3P1F}^{other} + \left[\frac{\epsilon(\ell_1)}{1 - \epsilon(\ell_1)} \frac{\epsilon(\ell_2)}{1 - \epsilon(\ell_2)} \right] \times N_{2P2F}^{data}. \quad (4.8)$$

A closure test of the pass-fail procedure has been performed on 2011 data in events with wrong flavour or charge combinations, where a very good agreement is observed between the predicted and observed yields, 11.6 and 12 events respectively.

Kinematic discriminator distribution. The distribution of the kinematic discriminator for reducible background events from the control region in data is compared to the ones extracted from simulated $Z + jets$ events and from simulated ZZ events. The three are found to be compatible, and the one from ZZ events is used as nominal distribution in the analysis as it benefits from a better statistical accuracy, with an associated systematical uncertainty determined from the difference of the three shapes.

Comparisons in the $Z_1 + \ell\ell'$ extended phase space. As an additional test of the modelling of the reducible background, data to simulation comparisons are performed also in an extended $Z_1 + \ell\ell'$ phase space obtained by requiring the presence of a Z_1 candidate plus two additional leptons of any flavour and charge satisfying the nominal lepton identification requirements, but no isolation and impact parameter requirements. All the kinematic requirements of the signal region are applied in this region. Within the limited statistical accuracy of the simulations, a good agreement is observed in general both for the lepton selection variables (Figure 4.9) and for the kinematics (Figure 4.10), although the simulation underpredicts to some extent the normalization of the reducible background.

4.5. Results

The observed four-lepton mass distribution is shown in Figure 4.11; for illustration purposes, the three final states and the two running periods are shown together, but they are treated separately in the analysis since the background levels and mass resolutions are different, and the uncertainties on the electron and muon energy scales are independent.

While the statistical analysis is performed only for $m_{ZZ} > 100 \, GeV/c^2$, the distributions are shown down to $70 \, GeV/c^2$: this allows to validate the search directly on the $Z \to 4\ell$ decay mode (Figure 4.12), with an event

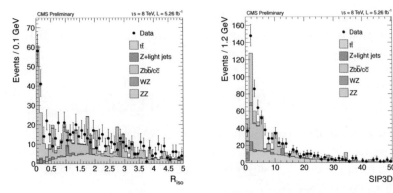

Figure 4.9. Data to simulation comparison of lepton observables in the $Z_1 + \ell\ell'$ phase space as defined in the text, in 2012 data. The relative isolation and the significance of the 3d impact parameter are shown for all the leptons in the left and right panel respectively. The signal region is defined by requirements $R_{\mathrm{iso}} < 0.4$, $SIP3D < 4$, as indicated by the red lines in the plots. In these plots, the prediction for all backgrounds are taken directly from simulations.

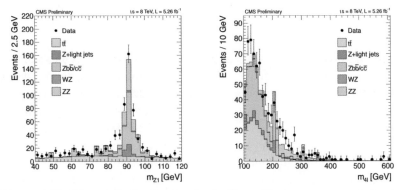

Figure 4.10. Kinematic distributions in the $Z_1 + \ell\ell'$ phase space as defined in the text, in 2012 data: m_{Z_1} on the left, m_{ZZ} on the right. In these plots, the prediction for all backgrounds are taken directly from simulations.

yield several times larger than the one from a light SM Higgs boson. The position and width of this known 4ℓ peak has also been used to validate the propagation of the lepton energy scales and resolutions from the $Z \to \ell^+\ell^-$ calibration samples to the softer leptons from $H \to ZZ \to 4\ell$ decays, although the tests are currently limited by the statistical uncertainty from the small number of $Z \to 4\ell$ events.

In the high mass region, the data are found to be fairly compatible with the expectations from the background. In the low mass region, instead, a localized excess of events compared to the background expectations is observed for m_{ZZ} close to $126 \, \mathrm{GeV}/c^2$. A comparison of the observed bi-

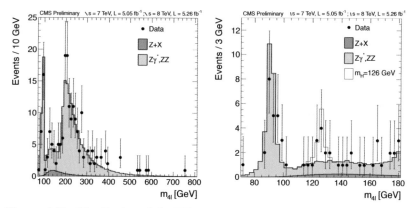

Figure 4.11. Distribution of the ZZ reconstructed mass, for the sum of the 4e, 4 μ, and 2e 2 μ channels in the combined 2011 and 2012 dataset. Points represent the data, filled histograms represent the expectations for the backgrounds. In the right plot, expanding the low mass range, the expectation for a SM Higgs boson of mass 126 GeV/c^2 is also shown as an outlined histogram.

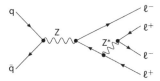

Figure 4.12. Feynman diagrams for the Z $\rightarrow$ 4ℓ decay process.

dimensional distribution in the mass distribution and kinematic discriminator are given in Figure 4.13, showing that the excess of events extends also to high, signal-like, values of the kinematic discriminator. This is also visible if the mass distribution is shown for the events satisfying a tight requirement $K_D > 0.5$ (Figure 4.14).

4.6. Searches in the H $\rightarrow$ ZZ $\rightarrow$ 2ℓ2τ decay channels

While the H $\rightarrow$ ZZ $\rightarrow$ 4ℓ decays with electrons and muons in the final state are clearly more accessible, the excellent tau reconstruction performance of CMS allows searches for a Higgs boson also in final states with taus. At present, the search is performed only above the 2 m_Z threshold, so that both Z bosons are on mass shell.

4.6.1. Event Selection

In the H $\rightarrow$ ZZ $\rightarrow$ 2ℓ2τ searches, events from the double-muon and double-electron trigger streams are first selected requiring the presence of a Z $\rightarrow$ e$^+$e$^-$ or Z $\rightarrow$ μ^+ μ^- candidate with 60 < $m_{\ell\ell}$ < 120 GeV/c^2;

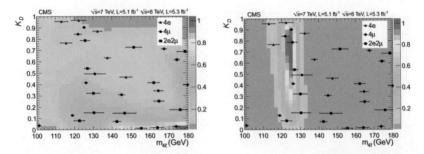

Figure 4.13. Bidimensional distribution of the invariant mass m_{ZZ} and the kinematic discriminator K_D, for the observed events and the predictions for the backgrounds (left) and an expected Higgs boson signal of mass $125\,\text{GeV}/c^2$ (right). The event-by-event uncertainties on the mass measurement of individual events are also indicated by horizontal error bars on each data point.

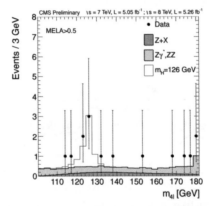

Figure 4.14. Invariant mass distribution of the observed events, and expectations from signal and background, after applying a requirement on the kinematic discriminator ($K_D > 0.5$).

the lepton identification criteria are very close to the ones used in the $H \to ZZ \to 4\ell$ searches, but the impact parameter requirements are relaxed when leptons from tau decays are considered.

A $Z \to \tau^+\tau^-$ candidate is then searched for in the four final states $e\,\mu$, $e\tau_h$, $\mu\tau_h$ and $\tau_h\tau_h$, where τ_h denotes hadronically decaying taus. The tau identification and isolation criteria used in the $\tau_h\tau_h$ selection correspond to single tau efficiencies of approximately 40% and a misidentification probabilities for hadronic jets of about 0.5%. In the $e\tau_h$ and $\mu\tau_h$ final states, a looser criteria corresponding to about 50% efficiency and 1% misidentification probability is used, compensated by tighter isolation requirements on the electrons and muons, as it is found to be more effective.

The invariant mass of the visible $Z \to \tau^+\tau^-$ decay products is required to be in the 30–90 GeV/c^2 range, to select mostly on-shell Z decays; the lower bound on the mass is removed in the e μ final state, characterized by a softer spectrum due to the larger number of neutrinos.

Expected yields. The expected signal event yield combining all H $\to$ ZZ $\to$ 2ℓ 2τ final states is about one order of magnitude smaller than the corresponding one for H $\to$ ZZ $\to$ 4ℓ decays with electrons and muons because of the tighter p_T and isolation criteria needed to suppress the larger backgrounds, and due to the lower reconstruction efficiency for hadronically decaying taus compared to the lighter leptons.

The main backgrounds to this search are the irreducible ZZ continuum production, and the Z+jets and WZ+jets processes with jets misidentified as leptons or hadronically decaying taus.

The expected signal and background yields are summarized in Table 4.3, together with the observed number of events, for the sum of the 2011 and 2012 LHC runs.

	all	$\tau_h \tau_h$	e τ_h	μ τ_h	e μ
signal	2.9 ± 0.7	0.67	0.73	0.92	0.59
ZZ continuum	12.1 ± 1.5	2.9	3.1	3.7	2.4
reducible bkg.	8.9 ± 2.5	3.6	2.5	1.5	1.3
observed events	20	4	9	2	5

Table 4.3. Expected and observed event yields after the full H $\to$ ZZ $\to$ 2ℓ2τ selection in each of the four Z $\to$ $\tau^+\tau^-$ final states considered, for SM Higgs boson signals of mass 200 GeV/c^2, the ZZ continuum and the reducible backgrounds.

4.6.2. Background estimation

As for the H $\to$ ZZ $\to$ 4ℓ search with electrons and muons, the largest background to the search is given by the electroweak ZZ production, which is estimated from simulations normalized to the NLO theoretical prediction for the cross section; correction factors for the lepton selection efficiencies are applied, measured with the tag-and-probe technique on Z $\to$ $\ell^+\ell^-$ and Z $\to$ $\tau^+\tau^-$ events.

The remaining backgrounds arise from Z + jets, W Z + jets, t$\bar{t}$ and to a lesser degree QCD multijet, and are all characterized by hadronic jets misidentified as taus or isolated leptons. These backgrounds are estimated with the method of misidentification probabilities, using events where an electron, muon or tau fails the isolation requirements.

The misidentification probabilities for electrons and muons are measured on the inclusive $Z + \ell$ events as for the $H \rightarrow ZZ \rightarrow 4\ell$ search. Misidentification probabilities for the isolation requirement applied to the τ_h candidates are measured from $2\ell\, 2\tau_h$ events in data for which the two tau candidates have the same electrical charge, which is dominated by $Z + jets$ events.

4.6.3. Results

In the data collected in the 2011 and 2012 LHC runs, a total of $6 + 14$ events is observed combining all $H \rightarrow ZZ \rightarrow 2\ell 2\tau$ final states, compared to an expected background of 21.0 ± 2.9 events.

The signal extraction in this search is performed from the invariant mass distribution of the visible $2\ell\, 2\tau$ decay products; the shapes of the expected distributions for signal and backgrounds are taken from simulations, and the background yields are normalized to the estimates described above. Due to the presence of undetected neutrinos in the tau decays, the mass resolution is 10–15%, and the peak of the expected distribution is at a value lower than the nominal Higgs boson mass by about 30%, as shown in Figure 4.15.

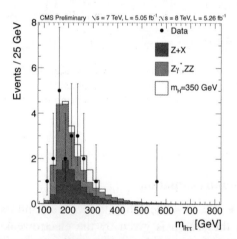

Figure 4.15. Reconstructed invariant mass distribution of the visible $2\ell\, 2\tau$ decay products, summed over the final states considered in the $H \rightarrow ZZ \rightarrow 2\ell 2\tau$ search. Points represent the data, filled histograms the expected background distributions and shaded histograms the expected signal distributions for two SM Higgs boson mass hypotheses.

Chapter 5
Searches for a light Higgs boson in the $\gamma\gamma$ and $\tau\tau$ final states

5.1. H → $\gamma\gamma$ channel

The H → $\gamma\gamma$ decay channel plays a major role in the searches for a Higgs boson in the low mass range. The branching ratio of the decay is only $O(0.1\%)$, but the diphoton final state provides a clean signature that can be exploited already at the trigger level, allowing an inclusive search to be performed. The overall strategy of the analysis is simple: a search for a narrow peak in the diphoton invariant mass spectrum over a smooth continuum background. In order to improve the performance, multivariate methods are extensively used to select events and categorize them in exclusive sets with different purity and mass resolution; an improvement of about 20% in sensitivity is achieved compared to more traditional methods used in the past [94].

5.1.1. Primary vertex identification

The H → $\gamma\gamma$ channel has the unique peculiarity that none of the Higgs boson decay products is detected in the silicon tracker, so that the determination of the primary vertex associated with the signal interaction is not trivial. It is nevertheless important to determine accurately the position of the interaction vertex, since it is needed to compute the opening angle between the two photons. The uncertainty on the opening angle gives a non negligible contribution to the diphoton mass resolution if the uncertainty on the z coordinate of the interaction is beyond 1 cm, and the RMS spread of the luminous region is about 6 cm.

Vertex identification algorithm. Vertex identification is based on three discriminating variables computed from the tracks associated to each vertex: an overall activity defined as the sum of the square of the momenta of the tracks $\sum p_T^2$, a recoil defined as the projection of the total vertex p_T along the direction identified by the diphoton momentum $\vec{p}_T(\text{vtx}) \cdot \hat{p}_T(\gamma\gamma)$, and an asymmetry between the total p_T of the vertex and the diphoton system $(p_T(\text{vtx}) - p_T(\gamma\gamma))/(p_T(\text{vtx}) + p_T(\gamma\gamma))$; for the last two definitions, $\vec{p}_T(\text{vtx})$ is taken to be the vector sum of the transverse momenta of the tracks in the vertex, and $p_T(\text{vtx}) = |\vec{p}_T(\text{vtx})|$ (Figure 5.1) The three variables are combined in a boosted decision tree trained on simulated H → γγ events.

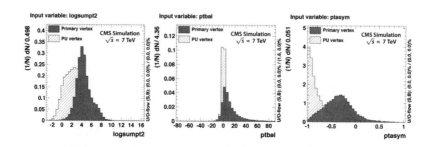

Figure 5.1. Distributions of the three input variables used for the multivariate primary vertex identification, for correct assignments (solid blue) and incorrect assignments (hatched red). From left to right: the logarithm of activity $\sum p_T^2$; the recoil or balance $\vec{p}_T(\text{vtx}) \cdot \hat{p}_T(\gamma\gamma)$; the asymmetry $|p_T(\text{vtx}) - p_T(\gamma\gamma)|/(p_T(\text{vtx}) + p_T(\gamma\gamma))$.

If at least one of the two photons converts in the tracker material and the conversion vertex is reconstructed, the vertex position along the z axis can also be estimated from the initial direction of the conversion track, or by extrapolating to the beam line the segment connecting the ECAL supercluster barycenter to the conversion vertex. These estimates are combined into a single determination z_{conv} with an associated uncertainty σ_{conv}, used to define for each vertex a pull variable $|z_{\text{vtx}} - z_{\text{conv}}|/\sigma_{\text{conv}}$. This pull variables are provided as an additional input to the boosted decision tree used for the vertex identification.

Performance. The overall vertex identification efficiency, expressed as the fraction of events in which the selected primary vertex is within 1 cm from the true one, is shown in Figure 5.2 as function of the diphoton p_T for the 2011 running conditions; due to the higher pile-up multiplicity in the 2012 run, the efficiency in that period is a few percent lower. The average efficiency for the full dataset, integrated over the expected p_T spectrum for a SM Higgs boson is about 80%.

The performance of the vertex identification for unconverted photon pairs is studied on data using Z → μ⁺ μ⁻ events, removing the two tracks of the muons from the inputs to the vertex reconstruction and identification algorithms, and using the dimuon momentum instead of the diphoton momentum when computing recoil and asymmetry. The measurement is repeated on simulated Z → μ⁺ μ⁻ events, and the two are found to be in good agreement except in the p_T region below 20 GeV/c, where the efficiency in data is 2–6% higher than in the simulation, possibly due to differences in description of the underlying event. As the p_T spectrum of H $\rightarrow \gamma\gamma$ events is different from the one of Z $\rightarrow \mu^+\mu^-$ events, the final estimate of the efficiency for the signal is obtained by re-weighting the simulated H $\rightarrow \gamma\gamma$ events using the ratio of the vertex identification efficiencies on Z $\rightarrow \mu^+\mu^-$ events between data and simulation in bins of the boson p_T.

The uncertainty on the vertex determination from converted photons σ_{conv} and the resulting identification efficiency is also measured on data, using balanced γ + jet events in which the tracks in the jet can be used to identify the correct primary vertex. The value of σ_{conv} depends strongly on the detector region where the conversion vertex is found, ranging from 0.02 cm for a conversion reconstructed in the pixel barrel to 0.9–1.2 cm for one reconstructed in the outer layers of the silicon tracker.

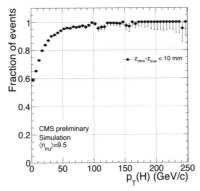

Figure 5.2. The fraction of events where the selected primary vertex is within 1 cm from the true one, for simulated H $\rightarrow \gamma\gamma$ events assuming a Higgs boson mass of 120 GeV/c^2.

5.1.2. Inclusive event selection

Events are selected by requiring two photons in the fiducial acceptance of the ECAL, excluding the transition region between barrel and endcaps. The photons are also required to be within the coverage of the silicon tracker, as track-based isolation and rejection of electrons are important to suppress the reducible backgrounds from hadronic jets.

Trigger. The data is collected using single and double L1 e/γ triggers with asymmetric E_T thresholds, and a combination of diphoton High Level triggers paths with two complementary photon selections, one based on calorimetric photon identification and loose isolation requirements, and one based only on the overall compactness of the electromagnetic shower. The E_T thresholds of the triggers are at least 10% lower than the ones used in the final selection, to minimize the turn-on effects taking into account also the different level of energy scale corrections that can be applied in the online selection with respect to the offline analysis. Electron rejection is not applied at trigger level, allowing the trigger performance to be studied with $Z \rightarrow e^+e^-$ events. The overall trigger efficiency for events passing the offline selection is greater than 99%.

Reconstruction.   Photon candidates are reconstructed offline from energy deposits in the ECAL, with the clustering algorithm described in Section 2.5.2. The offline thresholds applied to the two photons are function of the diphoton invariant mass $m_{\gamma\gamma}$, and are $p_T(1) > m_{\gamma\gamma}/3$ and $p_T(2) > m_{\gamma\gamma}/4$; this choice allows to reject more background at higher diphoton masses while preserving a good efficiencies for light Higgs boson masses.

In order to achieve the best energy resolution, after applying channel-by-channel calibration and transparency loss corrections, a multivariate regression method based on a boosted decision tree is used to correct for shower containment and energy losses in the material as function of the photon position and shower shape. A multivariate regression is also used to estimate the diphoton mass resolution for each event, used as input for the event categorization.

The energy scale calibration in data is extracted together with the energy resolution from a fit to the $Z \rightarrow e^+e^-$ lineshape, in bins R_9 and η; R_9 is a shower compactness variable defined as the fraction of the photon energy contained in a 3×3 matrix of ECAL crystals centered on the most energetic crystal, sensitive to whether photons interact in the detector material before the calorimeter. The overall uncertainty on the energy scale and resolution are similar, ranging between 0.1% and 0.9% of the photon energy, depending on the bin. The difference between the resolutions measured with this procedure on data and on simulated events is then applied as a smearing to the simulated signal events.

Identification and isolation.   The photon candidates are required to satisfy tight identification and isolation requirements in order to reject the backgrounds from $pp \rightarrow \gamma + jet$ and $pp \rightarrow jet + jet$ events.

The photon selection is based on a multivariate classifier implemented as a boosted decision tree relying on a combination of cluster shape and isolation variables, trained on simulated $H \rightarrow \gamma\gamma$ signal and $\gamma + jet$ background events; corrections determined from control samples in data are applied to the distributions of the input variables in simulation to improve the modelling. A loose preselection, including a veto for prompt electrons, is applied before the multivariate classifier.

Cluster shape variables include the longitudinal and transverse extension of the shower, R_9, and the ratio between the energy deposited in HCAL towers behind the ECAL supercluster and the energy of the supercluster.

In the analysis of the 2011 data, photon isolation variables are computed using tracks and energy deposits in the calorimeters in a cone around the photon, excluding those compatible with the footprint of the photon itself. Two techniques are used to mitigate the effects of pile-up on the performance of photon isolation: tracks which are not compatible with the selected primary vertex of the event along the z axis are discarded, and the calorimetric energy sums are corrected by a factor proportional to the per-event energy density. In addition to the combined isolation with respect to the selected primary vertex, two other variables are used: an isolation sum computed only from tracks, and a combined isolation using the primary vertex that maximizes the isolation sum; the latter is useful since the previous definition has less discriminating power for non-isolated photons coming from other collisions other than the selected one. In the 2012 data, the isolation variables are computed in a similar way but starting from the objects reconstructed from the particle flow algorithm.

In addition to the cluster shape and isolation variables, the multivariate classifier uses as input also the pseudorapidity of the photon and the number of reconstructed primary vertices in the event; these two variables allow the training to adjust the cluster shape and isolation requirements as function of the detector region and of the amount of pile-up.

The efficiency of the photon identification and isolation in each of the four categories is measured in data using $Z \to e^+e^-$ events, except for the electron rejection that is measured on $Z \to \mu^+\mu^-\gamma$ events. The latter events are also used to validate the agreement between the distributions of the identification variables in data and simulation. The values of the efficiencies measured in data and in simulation agree within 1–2% except for the low R_9 photons in the endcaps where the efficiency in data exceeds the one in the simulation by about 5%. For all the four categories, the ratio between the efficiencies in data and simulation is then applied as a weight to the simulated events.

Diphoton selection. A diphoton multivariate discriminator is used to classify the selected events into categories with different purity and mass resolution. The input variables to this second discriminator are: the transverse momenta of the two photons normalized to the diphoton mass;

the pseudorapidities of the photons; the cosine of the azimuthal angle between the photons; the relative diphoton mass resolutions computed under the two hypothesis of a correct or incorrect primary vertex identification, and the per-photon discriminating variables.

Also the diphoton classifier is implemented as boosted decision tree trained on simulated events. In the training, the optimization is performed correctly in terms of the purity in the narrow region compatible with a Higgs boson signal, and not on the full spectrum[1].

By design, this classifier is independent off the Higgs boson mass hypothesis chosen, and only smoothly dependent on the diphoton mass. A good agreement is observed between the output distribution of the diphoton discriminator in data and simulations when systematical uncertainties are taken into account for the modelling of the response of the single photon identification discriminator and the diphoton mass resolution estimator (Figure 5.3). The discriminator has also been validated in data using $Z \rightarrow e^+e^-$ events and diphoton events with $m_{\gamma\gamma} > 160 \, \text{GeV}/c^2$, dominated by true QCD diphoton production.

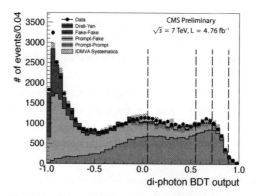

Figure 5.3. Distribution of the diphoton multivariate discriminator for data and simulations, for diphoton events in the 100–180 GeV/c^2 range. The systematical uncertainty from the modelling of the single photon identification discriminator is shown as a hatched area. The four vertical dashed lines correspond to the values used to define the diphoton categories used for signal extraction.

5.1.3. Selection of events with the VBF topology

The characteristic topology of VBF Higgs boson production, with two hadronic jets at large rapidities, allows to select a sample of diphoton events with a higher purity than the remaining, inclusive events.

---

[1] This is achieved by assigning to the events weights inversely proportional to the estimated diphoton mass resolution.

The selection used in 2011 data requires two jets, with p_T thresholds of 30/20 GeV/c, a large separation in pseudorapidity $\Delta\eta(jj) > 3.5$, and a large invariant mass $M(jj) \geq 350$ GeV/c^2. The p_T thresholds for the leading photons is tightened to $p_T(1) > 55/120 \times m_{\gamma\gamma}$, to better exploit the different p_T spectrum of VBF events.

In the analysis of 2012 data, to improve the sensitivity two classes of di-jet events are defined: a tight one, with the requirement $M(jj) \geq 500$ GeV/c^2 and $p_T(j) > 30$ GeV/c^2 for both jets, and a loose one with jet p_T thresholds of 30/20 GeV/c and an invariant mass requirement $M(jj) \geq 250$ GeV/c^2. The p_T threshold on the leading photon is defined to be equal to half the invariant mass of the diphoton pair.

In both running periods, two further selection requirements are applied: (i) the Zeppenfeld variable of the diphoton system [95], *i.e.* the difference between the average pseudorapidity of the two jets and the pseudorapidity of the diphoton system, is required to be less than 2.5; (ii) the dijet and diphoton systems are required to be back-to-back in the transverse plate, $\Delta\phi(\gamma\gamma, \mathrm{jj}) > 2.6$. The latter requirement provides no discrimination against genuine pp $\rightarrow$ $\gamma\gamma$ + 2jets, but allows to reject events in which any of the two jets come from a different pp interaction in the same bunch crossing.

The overall acceptance times efficiency for the 2011 VBF selection is 15% for Higgs boson events produced through the VBF channel, and 0.5% for events produced through the gluon fusion channel, for a Higgs boson mass of 120 GeV/c^2. However, since the gluon fusion cross section is an order of magnitude larger than the VBF cross section, the expected fraction of signal events from gluon fusion at the end of the selection is about 25% for the 2011 and tight 2012 selection, and about 50% for the loose 2012 selection.

5.1.4. Signal extraction

The last step of the search for a Higgs boson in the H $\rightarrow$ $\gamma\gamma$ channel is the analysis of the diphoton invariant mass spectrum, in the five or six event classes depending on the running period. The first four classes contain events not passing the VBF selection, categorized according to the output of the diphoton boosted decision tree; the number of classes and their boundaries have been obtained through an iterative optimization procedure using the expected exclusion limit as figure of merit. Events with very low values of the diphoton discriminator are discarded entirely, as they contribute negligibly to the sensitivity.

The additional one or two classes of photons contain the events that pass the VBF selections; also in these categories events with very low values of the diphoton discriminator are discarded, but since the yield in these classes is small, events are not further split in subclasses on the basis of the diphoton disciminator.

Background modelling.   In this search, the background shape and yield is extracted directly from the data: the only a-priori assumption made is that the background diphoton mass distribution can be well modelled by a smooth function.

The parametric forms chosen in the analysis in each category are polynomials of third to fifth order. While the choice of a high order polynomial is unphysical, it was proven to be the most robust choice when fitting Monte Carlo datasets generated assuming different background models. The possible systematic bias estimated from this procedure was found to be negligible with respect to the statistical uncertainties on the fitted background yield.

Signal modelling.   In each of the five diphoton categories, the signal is modelled as a superposition of Gaussians whose parameters were determined by fitting simulated signal events. The systematical uncertainties arising from the imperfect knowledge of the photon energy scales and resolutions in each category, and the vertex identification efficiency, which affects the tails of the resolution, are modelled as uncertainties on the parameters of the Gaussian mixture[2].

The parametric description of the model provides also a natural way to extend the model to hypothetical values of the Higgs boson mass for which simulated events are not available by smoothly interpolating between the values of the parameters at different points.

An alternative procedure of signal extraction has also been considered, based on the binned analysis of the output distribution of a final multivariate discriminator dependent on the Higgs boson mass instead of a parametric fit to the diphoton mass distribution. In this second model, the background shape and normalization are determined from multiple sidebands in the diphoton mass. The results obtained with the two approaches are found to be compatible.

Results.   The observed diphoton mass distribution after the final selection is shown in Figure 5.4 (left panel), including events from all the diphoton categories together, even if for the statistical analysis each category is considered separately. An alternative presentation of the same data is obtained by weighting the events in each category proportionally to the expected purity $S/(S + B)$, where S is the expected signal yield

---

[2] Simulated events with correct and incorrect vertex assignment are parameterized separately, and the vertex identification efficiency affects the relative fractions of the two events.

for a SM Higgs boson of mass $125\,\text{GeV}/c^2$ and B is the expected background in a narrow window around $m_{\gamma\gamma} = 125\,\text{GeV}/c^2$, so that the plot is no longer dominated by the categories with larger event yields and lower purity (Figure 5.4, right panel).

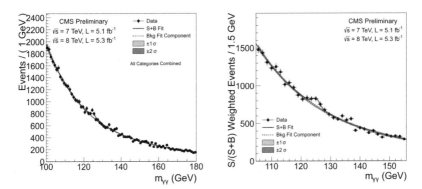

Figure 5.4. Observed diphoton mass distributions, for the simple sum of all diphoton categories (left) and for the weighted sum (right).

5.2. H → ττ channel

Thanks to the excellent tau reconstruction and trigger performances of CMS, the search for a Higgs boson in the ditau decay mode has been performed already on the data collected in the 2010 LHC run, in the context of the Minimal Supersymmetric Standard Model, where the cross sections and couplings between Higgs bosons and the taus can be greatly enhanced in some regions of the parameter space [96].

With the much larger integrated luminosity collected in 2011 and 2012, the H → ττ channel plays an important role also in the searches for a light SM Higgs boson, given the branching fraction of 5–8%. The search is performed using four ditau decay final states: $e\tau_h$, $\mu\tau_h$, $e\,\mu$ and $\mu\,\mu$, where τ_h denotes a tau decaying hadronically. In addition to lepton isolation and identification criteria, topological requirements involving the E_T^{miss} are used to improve the signal purity. A more complex selection is required in the $\mu\,\mu$ channel, to deal with the otherwise overwhelming Z → $\mu^+\mu^-$ background.

In order to improve the sensitivity, in all searches the events are separated into different categories according to the final state topology: (i) a VBF category, including events with two leading jets at large rapidity separation; (ii) a boosted category, in which at least one hadronic jet is required, selecting a topology with a better purity and a more precisely reconstructed ditau mass; (iii) a zero-jet category, characterized by larger event yields, which contributes to the sensitivity by constraining the $Z \to \tau^+\tau^-$ background yield. In the categories of events with zero or one jet, where a signal contribution is expected mostly from the gluon fusion production mode, events are further separated according to the p_T of the hadronic tau, or of the muon in the e μ and μ μ final state.

In the $e\tau_h$, $\mu\tau_h$, e μ final states, inference about the presence or absence of a Higgs boson signal is done from a binned analysis of the reconstructed ditau mass, while for the μ μ final state a two-dimensional binned analysis of the $(m_{\tau\tau}, m_{\mu\mu})$ is used due to the larger role of the $Z \to \mu^+\mu^-$ background.

5.2.1. Event selection

Trigger. Events are triggered requiring the presence of a $e\tau_h$, $\mu\tau_h$, e μ or μ μ pair; single muon triggers are also used for the search in the μ μ final state. While electrons and muons are already used to select events at the L1 trigger step, for these final states the hadronic taus are required only at the high level trigger step where the particle flow reconstruction can be applied, allowing for higher efficiencies especially at low tau momenta.

Lepton identification. The selection applied at the offline analysis step is based on tight identification and isolation criteria, in order to suppress reducible backgrounds from W + jets and multi-jet production. The electron and muon identification requirements are very similar to those used for the H $\to$ WW analysis, except that for muons the $\sigma(p_T)/p_T$ and the no kink selections are relaxed since mismeasured $Z \to \mu^+\mu^-$ events and $\pi^\pm/K^\pm$ decays are less of a concern.

Hadronic taus are reconstructed using the HPS algorithm, as described in Section 2.6.2. Muons misidentified as hadronic taus are rejected by requiring that the leading track of the decay is not reconstructed as a muon; in the $\mu\tau_h$ final state, τ_h candidates reconstructed as single hadrons are vetoed if the ratio between the energy deposited in the HCAL and the track momentum is below 0.2, to allow a further rejection of $Z \to \mu^+\mu^-$ events, independent of the muon system[3]. Electrons misidentified as taus are rejected using a multivariate classifier trained on simulated $Z \to e^+e^-$ and $Z \to \tau^+\tau^-$ events; the operation point chosen is tighter for the $e\tau_h$ final state, and looser for the $\mu\tau_h$ final state.

---

[3] Attempting to reject $Z \to \mu^+\mu^-$ events by requiring that the muon and leading track of the tau do not yield an invariant mass close to an on-shell Z boson would result in a significant efficiency loss for the signal.

Lepton isolation. For electrons and muons, isolation requirements are applied on the sum of the transverse energies of all the reconstructed particles in a cone around the lepton. To reduce the impact of pile-up on the isolation, charged particles not associated with the selected primary vertex are not included in the sum; the energy deposit from neutral particles is also corrected statistically by subtracting half the energy deposit from charged particles in the isolation cone associated to other primary vertices (the in-cone pile-up prescription described in Section 2.4.4).

For hadronic taus, a multivariate isolation discriminator is used, relying on the transverse energy sums in five concentric rings of $R = 0.1, \ldots, 0.5$, separately for charged hadrons, neutral hadrons and photons. Contamination from pile-up in the energy sums for neutral particles is subtracted using energy density corrections, and the median per-event energy density is also provided as an additional input to the discriminator. The training of the discriminator is done with simulated $Z \to \tau^+\tau^-$ events as signal and reconstructed tau candidates from multijet events in data as background.

Topological selection ($e\tau_h$, $\mu\tau_h$). The two leptons are required to be oppositely charged, where the charge of the hadronic tau is defined as the sum of the charges of its constituents. Due to the large ratio between the momenta considered and the tau mass, the neutrinos tend to be collinear with the other visible decay products of the tau; this allows to suppress the backgrounds from W + jets events in which a jet is misidentified as an hadronic tau, for which the neutrino tends to be back-to-back with the lepton from the W, by vetoing events with a large transverse mass of the lepton E_T^{miss} system.

Topological selection ($e\mu$). In the e μ channel, where the backgrounds W + jets are smaller with respect to those from $t\bar{t}$ and diboson production, a different topological requirement is applied: in the transverse plane, an axis ζ is identified as the bisector of the directions of the visible decay products of the two taus (Figure 5.5, left), and the projections along this axis of the visible ditau transverse momenta $p_\zeta^{\mathrm{vis}} = \vec{p}_T(\tau\tau) \cdot \vec{\zeta}$ and the missing transverse energy $p\!\!\!/_\zeta = \vec{E}_T^{\mathrm{miss}} \cdot \hat{\zeta}$ are computed. The two variables p_ζ^{vis} and $p\!\!\!/_\zeta$ are strongly correlated for signal and $Z \to \tau^+\tau^-$ events, but not for the other backgrounds, so good discrimination is achieved with a requirement on a linear combination of the two,

$$p\!\!\!/_\zeta - 0.85\, p_\zeta^{\mathrm{vis}} > -25 \text{ GeV}/c\,, \qquad (5.1)$$

as illustrated in Figure 5.5 (right).

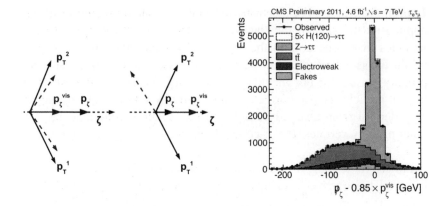

Figure 5.5. Left: prototypical $Z \to \tau^+ \tau^-$ or $H \to \tau\tau$ decay kinematic in the transverse plane, showing the transverse momenta of the visible and invisible decay products of each tau (solid and dashed arrows), the ζ axis, and the projections p_ζ and p_ζ^{vis} of the visible and total momentum of the tau along this axis. Middle: same representation for a typical background event where the total $\vec{E}_T^{miss}$ (dashed arrow) is unrelated to the momenta of the two tau candidates (solid arrows). Right: distribution of $p\!\!\!/_\zeta - 0.85\, p_\zeta^{vis}$ for e μ events in the $H \to \tau\tau$ analysis.

Topological selection ($\mu\mu$). In the μ μ final state, multivariate classifiers are used to reduce the $Z \to \tau^+ \tau^-$ and $Z \to \mu^+ \mu^-$ background. The discriminators are based on the following variables: the ratio of the transverse momentum of the dimuon system to the scalar sum of the two muon transverse momenta, $p_T(\mu\,\mu)/\sum_i p_T(\mu_i)$; the pseudorapidity of the dimuon system, $\eta(\mu\,\mu)$; the significance of the distance of closest approach between the two muon tracks, which is sensitive to the finite tau decay length; the azimuthal angle between the positive muon and the $\vec{E}_T^{miss}$ vector[4]. An additional binary variable is provided as input to the classifiers, according to whether a solution to the $H \to \tau\tau$ decay kinematic exists in the collinear approximation [97].

In the event categories with zero and one jet, the discriminator is implemented by means of a boosted decision tree, while in the VBF category where the event yields are smaller a simpler product of one dimensional likelihood ratios is used. Simulated signal and background events are used in the training of the multivariate classifier, but recoil corrections

---

[4] Due to the strong correlation between the azimuthal angles of the two muons for both signal and background, including also the $\Delta\phi(\mu^-, \vec{E}_T^{miss})$ does not improve the discrimination.

derived from $Z \to \mu^+ \mu^-$ events in data are used to improve the modelling of the E_T^{miss} response in the simulation.

5.2.2. Event classification

In the search for a SM Higgs boson, events are subdivided into five mutually exclusive event categories, to take advantage of the different Higgs boson production modes, the different kinematic distributions in between the signal and the irreducible $Z \to \tau^+ \tau^-$ background, and the better ditau mass resolution in events with larger tau transverse momenta.

VBF category.   Events with the vector boson fusion topology are selected requiring two tagging jets with $p_T > 30\,\text{GeV}/c$. To optimize the sensitivity for the VBF topology, the following seven kinematic variables are combined in a multivariate discriminator: the invariant mass and p_T of the dijet system, the $\Delta\eta$ and $\Delta\phi$ separation between the two jets, the p_T of the ditau system computed both including and not including the missing transverse energy, and the η separation between the visible part of the ditau system and the closest jet. In addition, a veto is applied on reconstructed jets with $p_T > 30\,\text{GeV}/c$ in the rapidity gap between the two tagging jets, and in the e μ final state also on b-tagged jets with $p_T > 20\,\text{GeV}/c$ independently of their rapidity.

The expected signal composition in this category is dominated by the VBF process, with a 20–30% contamination from gluon fusion process dependent on the final state considered.

Boosted category.   A boosted event category is defined by the requirement of at least one hadronic jet of $p_T > 30\,\text{GeV}/c$, excluding the events selected by the VBF category. These events are further subdivided into a low-p_T and high-p_T categories: for the channels with one hadronic tau, the high-p_T category is defined as $p_T(\tau_h) > 40\,\text{GeV}/c$; in the e μ final state the condition is $p_T(\mu) > 35\,\text{GeV}/c$, and in the μ μ final state it is $p_T(\mu) > 20\,\text{GeV}/c$. Events that contain any b-tagged jets with $p_T > 20\,\text{GeV}/c$ are vetoed.

The expected signal composition in this category is approximately 80% from gluon fusion production, the remainder being split among VBF and VH in a ratio of two to one. Despite the lower purity, the sensitivity of this category and the VBF one are very similar.

0-jet category.   The third category include events with no jets of p_T above $30\,\text{GeV}/c$. This last category contains the majority of events produced through the gluon fusion process, but also a very large irreducible background from $Z \to \tau^+ \tau^-$. The sensitivity of this category alone is about 50% worse than either the VBF or boosted categories, but its contribution to the combined analysis is significant as it provides constraints on the backgrounds, and also on the hadronic tau efficiency and energy scale in the $\mu\tau_h$ and $e\tau_h$ channels.

5.2.3. Background estimation

Irreducible Z $\rightarrow$ $\tau^+\tau^-$ background. The largest background in all event categories and final states except $\mu\mu$ is the electroweak $Z \rightarrow \tau^+\tau^-$ production. The prediction for this background is obtained using $Z \rightarrow \mu^+\mu^-$ events from data, in which the two muons are replaced with simulated taus of the same momenta. The overall yield is normalized using the measured $Z \rightarrow \mu^+\mu^-$ yield in data.

Irreducible Z $\rightarrow$ $\mu^+\mu^-$ background. For the $\mu\mu$ final state, the dominant background in all event categories is $Z \rightarrow \mu^+\mu^-$. The normalization for this background is derived directly from the data from a fit to the observed distribution of the significance of the distance of closest approach between the two muon tracks, separately in each event category and in different ranges of dimuon mass. The systematical uncertainty on this estimate is in the 4–6% range.

Reducible backgrounds. In the $e\mu$ channel, the background from events where a jet is misidentified as an electron are determined using the method of misidentification probabilities like in the $H \rightarrow WW$ analysis, and the much smaller rate of misidentified muons is neglected; this estimate includes QCD multi-jet events, W + jets events, and also the Z + jets events where one lepton is not reconstructed and a jet is misidentified as a lepton.

In the channels with hadronic taus, events from W + jets where the W decays to electron or muon and a jet is misidentified as an hadronic tau are measured in a control region defined by a large transverse mass of the lepton E_T^{miss} system, and extrapolated to the signal region using simulations.

The backgrounds from $Z \rightarrow \ell\ell$ events where a lepton is misidentified as an hadronic tau are estimated from the measured yield of well identified $Z \rightarrow \ell\ell$ events using misidentification probabilities measured in data with the tag-and-probe technique.

The remaining reducible backgrounds in the $e\tau_h$ and $\mu\tau_h$ channels, mostly from QCD multijet, are estimated from the events where the two leptons have the same charge, and extrapolated to the signal region using the ratio between opposite-charge and same-charge events measured in a control region obtained inverting the isolation requirement on the muon or electron. The contribution of events from W + jets in the same-charge sample is determined using events with large m_T, and subtracted before extrapolating to the signal region.

In the $\mu\,\mu$ channel, the reducible QCD multijet background is estimated from data using events where the two muons have the same charge. The reducible W + jets background, is estimated from simulations and found to be negligible.

Top and Diboson backgrounds.   The backgrounds from $t\bar{t}$ and diboson production are estimated from simulations. In the case of $t\bar{t}$, the inclusive cross section is normalized to the CMS measurement [98], with an 8% uncertainty; the background estimate is also checked in control regions dominated by $t\bar{t}$, where a good agreement is observed. Diboson cross sections are taken from NLO predictions, with a conservative 15% uncertainty.

5.2.4. Signal extraction

In this search, the signal is extracted from a binned likelihood fit to the observed ditau mass distribution in each final state and event category.

Ditau invariant mass reconstruction.   The invariant mass of the ditau system is reconstructed using a likelihood technique to estimate the momenta of the two neutrinos, taking into account the measured E_T^{miss} after recoil corrections, kinematic constraints, the expected tau momentum distribution as function of the ditau mass, and the tau-decay phase space. The algorithm yields a reconstructed invariant mass consistent with the true value, with nearly Gaussian shape. For a Higgs boson mass hypothesis of 130 GeV, the mass resolution is 21% for inclusive events, and 15% in the boosted and VBF categories where the ditau system is produced with a larger transverse momentum.

Results.   The observed ditau invariant mass distributions and the expectations from signal and backgrounds are shown in Figures 5.6 and 5.7, separately for each event category in the four decay final states. For illustration purposes the events from the two running periods have been combined in the same distribution, the p_T categorization in the zero-jet and boosted categories have likewise been merged, and in the $\mu\,\mu$ case the bidimensional $(m_{\tau\tau}, m_{\mu\,\mu})$ distributions have been projected to one dimension.

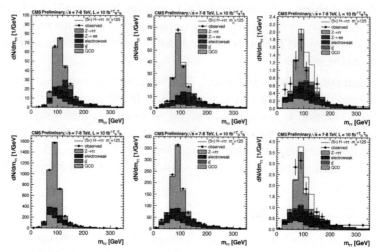

Figure 5.6. Observed and expected distributions of the ditau invariant mass for the combined 2011 and 2012 dataset in the different categories: 0-jet (left column), boosted (middle column), VBF (right column). The distributions in the upper row are for the $e\tau_h$ final state, while the ones in the bottom row are for $\mu\tau_h$. The low and high p_T sub-categories for the 0-jet and boosted final state are combined in these figures

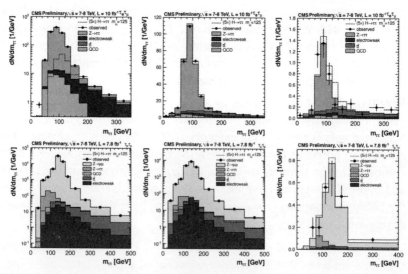

Figure 5.7. Observed and expected distributions of the ditau invariant mass for the combined 2011 and 2012 dataset in the different categories: 0-jet (left column), boosted (middle column), VBF (right column). The distributions in the upper row are for the $e\mu$ final state, while the ones in the bottom row are for $\mu\mu$. The low and high p_T sub-categories for the 0-jet and boosted final state are combined in these figures.

Chapter 6
Searches for VH and t̄tH associated production

6.1. VH, H → b̄b channel

The observation of the H → b̄b decay mode is extremely important in the context of Higgs boson searches, as it is a direct test the Yukawa couplings of the Higgs boson to quarks. Due to the overwhelming QCD multijet background in the gluon fusion and vector boson fusion production modes, the search is performed relying on associated VH production, with the vector boson decaying to leptons and/or neutrinos . In the CMS searches, five vector boson decay modes are considered: $W \to e\,\nu$, $W \to \mu\,\nu$, $Z \to ee$, $Z \to \mu\,\mu$, and $Z \to \nu\,\nu$, the latter is detected as a large unbalance in the transverse momentum of the event.

The main backgrounds in this search are the production of vector bosons in association with jets (especially $Wb\bar{b}$, $Zb\bar{b}$), single and pair production of top quarks, and diboson production. These backgrounds are substantially reduced by performing the search in the phase space region where the vector boson and the dijet system are produced with large transverse momenta in a back-to-back topology [99]. This kinematic regime is also characterized by a more accurate jet energy reconstruction, since the constituents are more collimated. In the analysis of each channel, two event categories are defined respectively for moderate and for large transverse momenta of the vector boson, the precise requirements depending on the specific final state. To further improve the sensitivity, a multivariate analysis technique is used: a discriminator is trained on simulated signal and background events for a number of Higgs boson masses, and a binned analysis of the output distribution of the discriminator is performed, as in the H → WW → $2\ell2\,\nu$ analysis.

6.1.1. Event selection

In the trigger, events are selected relying on the presence of one or two leptons, or missing transverse energy in the $Z \to \nu\nu$ case. For the

W → e ν case, at large instantaneous luminosities an acceptable trigger rate is achieved by the combination two selections, one with a large p_T threshold on the electron and another with a lower threshold but complemented by the request of two jets and a moderate amount of missing transverse energy. Likewise, in the Z → $\nu\nu$ case triggers requiring the presence of one or two central jets within the tracker coverage and looser E_T^{miss} requirements are deployed in addition to one solely based on a tight E_T^{miss} requirement. For both jets and missing energy, particle-flow reconstruction is used in the high level trigger. The trigger efficiencies for events satisfying the offline selection requirements are above about 98% for ZH modes, and slightly less for the W → $\ell\nu$ cases, 90% for muons and 95% for electrons.

Offline, choices similar to that of the other Higgs analyses are used for leptons, jets and missing transverse energy. Particle flow reconstruction is used consistently as input for jet reconstruction, using the anti-k_T clustering with distance parameter $R = 0.5$, lepton isolation criteria and missing energy determination; for the lepton isolation and jet reconstruction purposes, charged particles associated to pile-up primary vertices are discarded, and the residual contamination from neutral particles is subtracted using a median per-event energy density. For this analysis, the combined secondary vertex b-tagger is used, as it provides the best performance in the kinematic regime considered; depending on the final state, different working points are considered; the tightest one corresponds approximately to a b-jet efficiency of 50% with 6% and 0.15% efficiencies c-jets and light jets respectively while the loosest one is associated to about 82% efficiency for b-jets and 12% efficiency for light jets.

Identification of W → $\mu\nu$ and W → e ν decays is done requiring the presence of a single muon with $p_T > 20$ GeV/c or a single electron with $p_T > 30$ GeV/c, and no other isolated leptons with $p_T > 20$; in the electron case, the additional requirement $E_T^{miss} > 35$ GeV/c^2 is enforced to suppress QCD-induced backgrounds. The transverse momentum of the W boson is obtained from the vector sum of the lepton $\vec{p}_T$ and the $\vec{E}_T^{miss}$. Two event categories are defined on the basis of the p_T of the W: $120 < p_T(W) \leq 170$ GeV/c, and $p_T(W) > 170$ GeV/c.

Z → $\ell^+\ell^-$ decays are identified by the presence of two leptons with $p_T > 20$ GeV/c, with the same flavour but opposite electrical charge, and an invariant mass in the 75–105 GeV/c^2 range. As for the W case, two p_T categories are defined, with requirements looser than in the W case to compensate for the smaller signal cross section times branching ratio for ZH → $\ell\ell b\bar{b}$ compared to WH → $\ell\nu b\bar{b}$: $50 < p_T(Z) \leq 100$ GeV/c, and $p_T(Z) > 100$ GeV/c

For Z $\rightarrow \nu\nu$ decays, the transverse momentum of the Z boson is equal to the missing transverse momentum, and thus two event categories are defined as $120 < E_T^{\mathrm{miss}} \leq 160\,\mathrm{GeV}$ and $E_T^{\mathrm{miss}} > 160\,\mathrm{GeV}$. To suppress the backgrounds originating from instrumental missing energy, an azimuthal separation $\Delta\phi > 0.5$ radiants is required between the E_T^{miss} and the closest jet with $p_T > 20\,\mathrm{GeV}/c$, $|\eta| < 2.5$. A veto on isolated leptons with $p_T > 20\,\mathrm{GeV}/c$ is also applied.

H $\rightarrow$ b$\bar{\mathrm{b}}$ decay candidates are formed from pairs of b-tagged jets with $|\eta| < 2.5$ and a p_T requirement dependent on the final state: 30 GeV/c for WH, 20 GeV/c for leptonic ZH, and an asymmetric 80/20 GeV/c requirement for invisible ZH enforced by the jet requirements applied at trigger level. In case of multiple candidates, the pair with the largest dijet p_T is selected. A transverse momentum requirement is applied to the dijet system, equal to that applied to the vector boson.

In order to improve the mass resolution for the dijet pair, a multivariate regression algorithm based on a boosted decision tree is used, combining kinematic variables and detailed information about the jet structure and the products of the b decay candidate, *e.g.* charged hadron tracks and secondary vertices. The regression is trained on simulated events, but its performance is validated in data from the p_T balance in Z + b$\bar{\mathrm{b}}$ events (fig. 6.1). The improvement in the mass resolution from this method is about 15%, resulting in a gain in sensitivity for the overall analysis of 10–20%, depending on the channel.

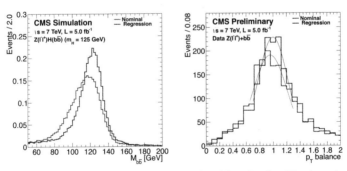

Figure 6.1. Performance of the regression algorithm for the dijet invariant mass on simulated signal events for $m_{\mathrm{H}} = 125\,\mathrm{GeV}/c^2$ (left), and validation on Z+b$\bar{\mathrm{b}}$ events in 2011 data.

6.1.2. Background estimation and systematical uncertainties

A good control of the backgrounds is very important in this analysis, as contrary to the H $\rightarrow \gamma\gamma$ case the discriminating power offered by the reconstructed mass peak is not enough to obtain a good signal sensitivity without any prior knowledge of the background.

Because of this, the normalization of the dominant V + jets and $t\bar{t}$ backgrounds is determined from control regions in the data. Separate control regions are used in each final state for vector boson production in association with light jets and with heavy flavour jets, to avoid relying on the theoretical predictions on the ratio between the two cross sections. The control regions are defined to be exclusive, and the data to simulation scale factors for the three backgrounds are determined simultaneously from all control regions at the same time; this approach is necessary since contaminations from other backgrounds in the control regions are sizable.

Systematical uncertainties on the background determination from control regions is assessed through the variation of the estimate when the selection criteria defining the control regions are varied around their nominal values. The overall statistical plus systematical uncertainties on these backgrounds are in the 10%–30% range depending on the final state and background considered.

The background from QCD multijet production in the $Z \rightarrow \nu\nu$ channel is estimated from data by means of a two dimensional extrapolation using the sum of the b-tagging discriminators of the two jets and the $\Delta\phi$ between the $\vec{E}_T^{\text{miss}}$ and the closest jet. The estimated yield is found to be negligible.

The expected yields for the signal and for the remaining backgrounds, *i.e.* diboson and single top production, are estimated from simulations. Systematical uncertainties are assigned for the normalization of the integrated luminosities, the b-tagging efficiencies, the production cross sections and the extrapolation to the high p_T regime. For the last source of uncertainty, a 30% was assumed in the diboson and single top cases for which no dedicated theoretical evaluation of the uncertainty is available; the corresponding uncertainty for signal is about 15%, so this estimate can be considered conservative.

6.1.3. Signal extraction

The extraction of a Higgs boson signal from the sample of events satisfying the selection described above relies on multivariate methods. A boosted decision tree (BDT) is trained on simulated signal and background events, relying on kinematic and angular variables, the b-tagging discriminators of the two jets, the number of additional central jets, and a color pull variable sensitive to the hadronization pattern for the color-singlet H $\rightarrow$ b$\bar{b}$ decay [100]. For this purpose, the differences in b-tagging performance between data and simulation are taken into account by applying a continuous deformation to the distribution of the b-tagging discriminators in simulated signal and background events to reproduce the efficiencies and mistag probabilities measured in data; this is anyway a small correction, since simulation and data agree to 10% or better both in the efficiencies and in the mistag rates.

Separately for the $\sqrt{s} = 7\,\text{TeV}$ and $8\,\text{TeV}$ periods, boosted decision trees are trained individually for each of the five final states, and for six different Higgs boson mass hypotheses in the range 110–$135\,\text{GeV}/c^2$. A binned likelihood analysis is performed on the output distribution of the discriminator, using templates for the signal and backgrounds extracted from simulations. The background templates are normalized according to estimates of the background yields from control regions in data, and both normalization and shape uncertainties for the signal and background templates are considered in the procedure.

In general, a fair agreement is found between the observed distribution of the BDT classifiers in data and the expectations from the background. Prototypical distributions for two of the channels in the 2012 data are displayed in Figure 6.2.

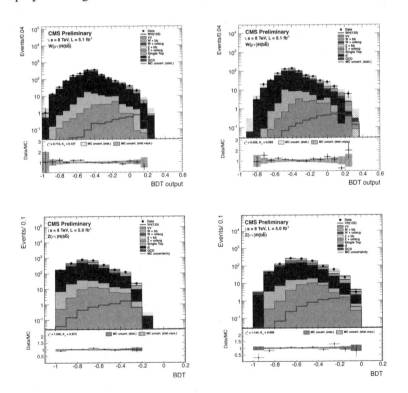

Figure 6.2. Distributions of the BDT outputs for the $m_\text{H} = 125\,\text{GeV}/c^2$ hypothesis for the 2012 data in the $W \rightarrow \mu\,\nu$ channel (top row) and $Z \rightarrow \nu\,\nu$ channel (bottom row), for the low and high $p_T(V)$ categories.

For illustration, the inclusive dijet invariant mass distribution for the events from all final states and running periods together is shown in Figure 6.3. A tighter event selection is used in this context, optimized for the best sensitivity if the dijet invariant mass were used as final discriminating variable in the search.

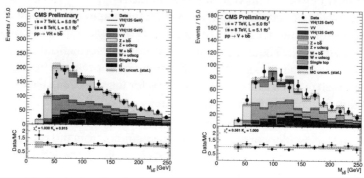

Figure 6.3. Inclusive distributions of the dijet invariant mass for the combination of 7 TeV and 8 TeV data in all final states together, for the low and high $p_T(V)$ categories, for events passing a selection optimized for this variable. The contributions from signal and diboson background are shown also as non-stacked hollow histograms.

6.2. WH → $3\ell3\nu$ channel

The WH → $3\ell3\,\nu$ search is a natural extension to the H → WW → $2\ell2\,\nu$ search to trilepton final states. While the sensitivity is drastically reduced by the smaller cross section times branching ratio, the complementarity of the production mode is useful when testing for Higgs boson scenarios with couplings different from the standard model ones.

For this search, currently only results from the $\sqrt{s} = 7$ TeV running are available from CMS.

Event selection. This analysis relies on the same physics objects used in the H → WW → $2\ell2\,\nu$ analysis (Section 3.3). The starting point of the selection is the requirement of exactly three identified and isolated leptons, with total charge ±1. The leading lepton is required to have $p_T > 20$ GeV/c, and the other two $p_T > 10$ GeV/c.

The minimum of the projected missing energies computed from all particle flow objects (E_T^{miss}) and only from charged ones ($E_{T,\mathrm{trk}}^{\mathrm{miss}}$) is required to be larger than 40 GeV, suppressing backgrounds with no genuine E_T^{miss} such as Z + jets, Z + γ and ZZ. To further reject events containing Z → $\ell^+\ell^-$ decays, e.g. from WZ → $3\ell\,\nu$, all pairs of same-flavour opposite-charge leptons are required to have invariant mass at least $25\,\mathrm{GeV}/c^2$ away from the nominal Z boson mass. To control the backgrounds from $t\bar{t}$ and tW production, events are rejected if there is a jet with $p_T > 40\,\mathrm{GeV}/c$, a b-tagged jet with $p_T > 10\,\mathrm{GeV}/c$, or a soft muon. Backgrounds from low mass resonances and photon conversions are also suppressed requiring all three dilepton invariant masses to be larger than $12\,\mathrm{GeV}/c^2$.

As the search is aimed at light Higgs boson mass hypotheses, the smallest of the three dilepton masses is required to be below $100\,\mathrm{GeV}/c^2$, and the ΔR between the the closest pair of leptons is required to be smaller than 2.

The expected background yield after the full selection is about 8 events, two thirds of which arise from $WZ \to 3\ell\,\nu$ and the remaining dominated by reducible backgrounds (dominantly $t\bar{t}$ and tW, with some Z+jets). The expected signal yield for a Higgs boson mass hypothesis of $120\,\mathrm{GeV}/c^2$ is 0.6 events, 80% of which from $WH \to 3W \to 3\ell 3\,\nu$ and the remaining from $WH \to W\tau\tau \to 3\ell 5\,\nu$.

Background estimation.   The dominant $WZ \to 3\ell\,\nu$ background is estimated from data in an orthogonal trilepton control sample where a pair of same-flavour opposite-charge leptons has mass within $15\,\mathrm{GeV}/c^2$ from that of an on-shell Z boson; in the definition of the control sample, the E_T^{miss} requirement and the jet and b-jet vetoes are preserved, while the requirements on the ΔR and invariant masses of the leptons are relaxed. The extrapolation to the signal phase space is done on the basis of simulated events. An uncertainty of 12% is assigned to the normalization of this background, driven by the statistical uncertainty from the limited number of events in the control sample.

The reducible backgrounds are also estimated from data, using the method of the misidentification probabilities starting from events in which one of the three leptons fails the isolation or identification requirements, similarly to what done for the W + jets background in the $H \to WW \to 2\ell 2\,\nu$ search. The overall uncertainty on this estimate is about 30%.

Simulated events are used to estimate the remaining small backgrounds from $ZZ \to 4\ell, Z + \gamma$, WWW and other rare processes such as $t\bar{t}W$ and $t\bar{t}Z$.

Signal extraction.   Due to the very small expected signal and background yield, and the lack of a clear reconstructed mass peak the analysis is performed as a counting experiment, with all lepton flavours considered together. The signal prediction is taken from simulated events, and the searches for different Higgs boson mass hypotheses only differ in the expected signal yield considered when interpreting the observed results.

6.3. WH → τₕ2ℓ channel

The WH → $\tau_h 2\ell$ search is performed as an extension to the H → ττ search to cover the associated production mode, relying on the additional charged lepton from the W boson decay to improve the background rejection. Two final states are considered, τ_h e μ and τ_h μ μ; the τ_h ee final state is not used due to the larger reducible backgrounds.

As for the WH → $3\ell 3\,\nu$ search, also for this search at the moment only the data from the $\sqrt{s} = 7$ TeV has been analyzed.

Event selection. This analysis relies on the same physics objects as the H → ττ search (Section 5.2). Events are selected requiring a hadronically decaying tau candidate with $p_T > 20$ GeV/c and $|\eta| < 2.3$, a muon with $p_T > 20$ GeV/c and $|\eta| < 2.1$, and a second muon or electron $p_T > 10$ GeV/c and $|\eta| < 2.1/2.5$ for muons and electrons respectively. Compared to the H → ττ search, the isolation requirements on electrons and muons are loosened to cope with the smaller production cross section.

The sum of the electrical charges of the three leptons is required to be ±1, and the two light leptons are required to have the same electrical charge, in order to suppress contaminations from background processes with two light leptons and a jet misidentified as tau, *e.g.* from Z → $\mu^+ \mu^-$, WW, $t\bar{t}$, tW and Z → $\tau^+\tau^- \to e^\pm \mu^\mp 4\,\nu$.

Events that include additional isolated electrons or muons with $p_T > 10$, or a b-tagged jet with $p_T > 20$ GeV/c are rejected, to suppress contaminations from ZZ, $t\bar{t}$ and tW production.

Additional discrimination against reducible backgrounds is achieved by the requirement that the sum of the transverse energies of the three lepton candidates in the event L_T be larger than 80 GeV/c (fig. 6.4).

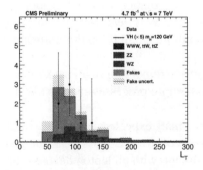 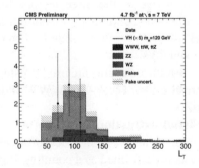

Figure 6.4. Distributions of the sum of the transverse energies of the three leptons L_T after the full selection except for the L_T requirement. The plots on the left and right panels refer to the τ_h μ μ and τ_he μ final states, respectively.

Background estimation. The irreducible backgrounds from WZ, ZZ, WWW and $t\bar{t}$W are estimated from simulations; for diboson production, the cross sections are normalized to the NLO predictions, while for the remaining rare processes leading order calculations from MADGRAPH are used, with a 100% systematical uncertainty. The contribution of WZ to the total background yield is about 40% in the τ_h μ μ final state and about 25% in the τ_h e μ one, while all other irreducible backgrounds together contribute to about 5% of the total background yield.

The requirement that the two light leptons have the same electrical charge implies that all reducible backgrounds are from events with a misidentified electron or muon, *e.g.* from Z $\rightarrow$ $\tau^+\tau^-$ or W+jets, with the exception of small contributions from WZ or ZZ with a misidentified τ_h that are already included in the irreducible background estimate. The estimation of these reducible backgrounds is performed using the method of misidentification probabilities, starting from control samples where one or both of the light leptons fail the isolation requirements. The overall uncertainty on the normalization of the reducible backgrounds amounts to approximately 30% in both final states.

Signal extraction. As in the WH $\rightarrow$ $3\ell\,3\,\nu$ case, due to the small event yield and the lack of a precise mass reconstruction the search is performed as a counting experiment, but in this case the two final states are kept separate. In total, 5 τ_h μ μ and 4 τ_he μ events are observed in data, compatible with the background expectations of 5.6 and 8.2 events respectively. The expected signal yield for a SM Higgs boson of mass 125 GeV/c^2 is about 0.4 events in each final states, mainly from H $\rightarrow$ $\tau\tau$ decays but with a 20% contribution from H $\rightarrow$ WW with one W boson decaying to τ_h ν.

6.4. $t\bar{t}$H, H $\rightarrow$ b$\bar{b}$ channel

The associated production of a Higgs boson with a top anti-top pair, $t\bar{t}$H, provides the only direct measurement of the Yukawa coupling of the top quark for Higgs boson masses lighter than the $t\bar{t}$ threshold. This measurement is therefore very important when combined with the same measurement extracted from the gluon fusion production cross section, as it can constrain the contribution of possible beyond-standard-model particles in the loop.

Because of the small cross section, the search can be performed only in the H $\rightarrow$ b$\bar{b}$ decay mode, and the dominant background is $t\bar{t}$ production with associated jets.

Event selection. In this search, events are selected requiring at least one of the top quarks to decay leptonically t $\rightarrow$ ℓ νb.

Jets are reconstructed using the anti-k_T algorithm with distance parameter $R = 0.5$, from the output of the particle flow algorithm; charged hadron candidates not associated to the primary vertex are removed prior to the jet clustering.

In the final states with a single lepton, $t\bar{t} \rightarrow \ell \, \nu q \bar{q}' b \bar{b}$, a lepton with $p_T > 30 \, \text{GeV}/c$ and satisfying tight identification and isolation criteria is required, accompanied by at least four hadronic jets in the central part of the detector $|\eta| < 2.4$, three with $p_T > 40 \, \text{GeV}/c$ and a fourth with $p_T > 30 \, \text{GeV}/c$.

In the dileptonic final states, $t\bar{t} \rightarrow 2\ell 2 \, \nu b \bar{b}$, the selection requirements are one tight lepton with $p_T > 20 \, \text{GeV}/c$, one loose lepton with $p_T > 10 \, \text{GeV}/c$ and at least two hadronic jets with $|\eta| < 2.4$ and $p_T > 30 \, \text{GeV}/c$.

In both final states, at least two of the hadronic jets have to be b-tagged using the combined secondary vertex algorithm (Section 2.3.6), with an operating point corresponding to an efficiency of 70% for jets originating from a b quark and a mistag probability of 2% for light quark or gluon jets.

Signal extraction. Events are categorized according to the multiplicity of hadronic jets and b-tags, into a total of 9 exclusive final states as listed in table 6.1. To discriminate between signal and background in each category, a multivariate discriminator based on an artificial neural network is trained using kinematic, b-tagging and angular variables, *e.g.* the minimal ΔR separation between two b-jets. In the dileptonic final state with at least three jets, the number of jets is also used as a discriminating variable. While the signal purity in some of the higher statistics channels

leptons	jets	b-tags	t$\bar{\text{t}}$H	t$\bar{\text{t}}$	others	data
1ℓ	4	3	3.5	981.6	60.1	1214
1ℓ	4	≥ 4	0.5	18.6	1.4	18
1ℓ	5	3	4.7	637.3	29.6	736
1ℓ	5	≥ 4	1.2	30.8	1.0	37
1ℓ	≥ 6	2	6.3	2160.3	95.4	2137
1ℓ	≥ 6	3	4.4	391.0	13.9	413
1ℓ	≥ 6	≥ 4	1.7	38.4	1.0	49
2ℓ	2	2	0.7	3354.1	951.9	4401
2ℓ	≥ 3	≥ 3	2.9	164.3	27.7	192

Table 6.1. Expected and observed event yields in each category of the $t\bar{t}$H analysis. The yields in the $t\bar{t}$ column are inclusive of the $t\bar{t} + c\bar{c}$, $t\bar{t} + b\bar{b}$ and $t\bar{t} + W/Z$ processes, while the others contains contributions from single top production and other non-top electroweak processes (W + jets, Z + jets, diboson). The expected signal yields are for a Higgs boson mass hypothesis of $120 \, \text{GeV}/c^2$.

is very low, they still contribute to the overall sensitivity by constraining the background in the other final states through correlations.

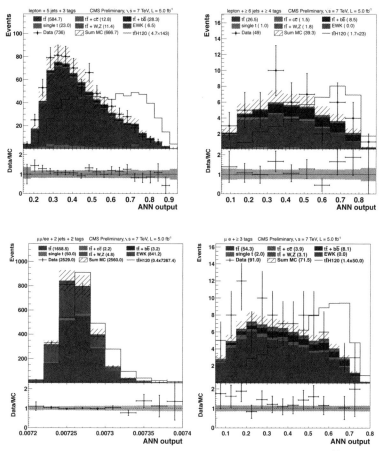

Figure 6.5. Distribution of the multivariate discriminator for the $t\bar{t}H$ search in four final states: $1\ell + 5$jets (3b), $1\ell + \geq 6$jets (≥ 4b), $\ell\ell + 2$b (ee and $\mu\mu$ combined), and e $\mu + \geq 3$b. The signal distribution is rescaled to the same yield as the total background prediction. The statistical and systematical uncertainty on backgrounds are shown by the hatched area.

Chapter 7
Searches for a heavy Higgs boson decaying into WW, ZZ in final states with hadronic jets and neutrinos

7.1. $H \rightarrow ZZ \rightarrow 2\ell 2q$ channel

The $H \rightarrow ZZ \rightarrow 2\ell 2q$ decay mode has the largest branching fraction among all $H \rightarrow ZZ$ modes considered at CMS, about 20 times larger than that of the fully leptonic final state, and since all four decay products are detectable the kinematic of the event can be fully reconstructed.

The main challenge to the searches in this channel is the background from the $Z + 2$ jets process, whose production cross section is four orders of magnitude larger than the signal one. Kinematic quantities such as the dijet invariant mass and the angular spin correlations of the decay products are used to discriminate between the signal and the background. Further improvement to the sensitivity is achieved by classifying the events according to the probability for the jets to originate from gluons, light quark or heavy quarks, exploiting the different flavour composition of the jets from $Z \rightarrow 2q$ decays with respect to that of the two additional jets in $Z + 2$ jets.

After the selection and categorization of the events, a signal is searched for in the invariant mass spectrum of the four reconstructed decay products; the signal model is obtained from simulations, while the normalization and shape of the $Z+2$jets background is determined from sidebands in the dijet mass. The search is performed separately for the two Higgs boson mass ranges 130–170 GeV/c^2 and 200–600 GeV/c^2, thereafter referred as low-mass and high-mass regions, but only in the latter the sensitivity is competitive with the one from the other searches.

This search has so far been performed only on the data collected at $\sqrt{s} = 7$ TeV, although an update to the full dataset is foreseen for the future.

7.1.1. Event selection in the high-mass region

In this analysis, events collected using the double electron, double muon and single muon triggers are used.

Events are selected offline requiring the presence of two identified and isolated leptons with same flavour and opposite electrical charge, and

p_T above 40 and 20 GeV/c for the leading and subleading lepton. Since both the signal and the main backgrounds are characterized by two real leptons from a Z decay, this analysis is not very sensitive to the choice of the isolation and identification criteria.

Jets are reconstructed with the particle flow, using the anti-k_T algorithm with distance parameter $R = 0.5$, and the the contamination from pile-up to the jet energies is subtracted using the median energy density computed from the FASTJET algorithm. To maximally benefit from the particle flow technique, only jets within the tracker acceptance $|\eta| < 2.4$ are considered. Jets are required to have $p_T > 30$ GeV/c, and a very loose selection is applied on the fraction of energies in the different subdetectors in order to suppress jets from calorimetric noise. Jets that overlap with isolated leptons within $\Delta R = 0.5$ are discarded.

Kinematic selection.   For Higgs boson masses above $2m_Z$, both Z bosons are expected to be on mass shell. The reconstructed dijet and dilepton invariant mass are therefore required to satisfy $75 < m_{jj} < 105$ GeV/c^2 and $70 < m_{\ell\ell} < 110$ GeV/c^2; the use of a narrower window for the dijet system, corresponding to about $\pm 2\sigma$, is motivated by the larger yield of the Z + 2 jets background compared to the non-Z backgrounds like $t\bar{t}$. After this requirement is applied, the kinematics of the two jets is refitted imposing the kinematic constraint $m_{jj} = m_Z$, taking into account the uncertainties on the jet momenta and directions. This fit improves significantly the resolution on the four-body mass m_{ZZ} also removes the correlation between m_{jj} and m_{ZZ}, simplifying the definition of signal and sideband region.

Angular likelihood discriminator.   Since the Higgs boson has spin zero, the angular distribution of its decay products is independent of the production mechanism, and can be parametrized in terms of five angles θ^*, Φ_1, θ_1, θ_2, Φ (Figure 7.1) which are mostly uncorrelated with the masses of the two Z bosons [87]. A likelihood discriminant is built from the joint probability in the five angles under the signal and background hypothesis $P_{\text{sig}}/(P_{\text{sig}} + P_{\text{bkg}})$.

The signal probability density P_{sig} is built from a correlated five-dimensional parametrization from the tree-level matrix element, corrected for the non-uniform detector efficiencies using empirical polynomial acceptance functions determined from simulation. The probability density P_{bkg} for the Z + 2 jets background is taken as product of one-dimensional distributions for each angle extracted from simulated events (Figure 7.2).

Events are selected by requiring the likelihood discriminator to be above a given threshold optimized depending on the Higgs boson mass hypothesis: beyond that, neither the five-dimensional probability densities nor the distribution of the likelihood discriminator are used in the signal extraction. This makes the result more robust, since this way the effect of the neglected correlations between the five angles or possible mismodelling of the backgrounds in the simulation cannot result in a bias but only in a slight loss of performance.

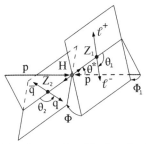

Figure 7.1. Definition of the five angles θ^*, Φ_1, θ_1, θ_2, Φ for the pp $\rightarrow$ H $\rightarrow$ ZZ $\rightarrow$ $2\ell 2$q process.

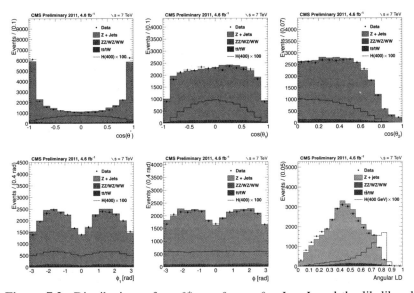

Figure 7.2. Distributions of $\cos\theta^*$, $\cos\theta_1$, $\cos\theta_2$, Φ_1, Φ and the likelihood discriminator $P_{\mathrm{sig}}/(P_{\mathrm{sig}} + P_{\mathrm{bkg}})$. Points with error bars show distributions of data, solid histograms depict the background expectation from simulated events with the different components illustrated. Open histograms indicate the expected distribution for a Higgs boson with mass $400\,\mathrm{GeV}/c^2$, multiplied by a factor of 100 for illustration.

Jet flavour categorization. The flavour composition of jets from hadronic Z decays is characterized by a large fraction of $b\bar{b}$ (22%) and $c\bar{c}$

(17%) decays, and is well known from LEP and SLD measurements [54]. Conversely, the $Z + 2$jet events are expected to contain a large contribution from gluon-induced jets, and a flavour composition of quark jets reflecting the proton parton density function with larger fractions of u and d quarks. In order to exploit these differences, events are categorized in three classes using the track counting high efficiency b-tagging algorithm (Section 2.3.6): events with both jets tagged as b-jets, events with only one tagged jet, and events with no b-jets.

A fourth class of gluon-tagged jets is extracted from the class of events with no b-jets using a quark-gluon discriminator built from three observables related to the jet fragmentation: the number of reconstructed charged hadrons, the number of reconstructed photons and neutral hadrons, and a variable $PTD = \sqrt{\sum p_T^2 / (\sum p_T)^2}$. The PTD is related to the fragmentation $z = p_T(\text{particle})/p_T(\text{jet})$, and is approximately equal to $\sqrt{\sum z^2}$. Jets induced by gluons are characterized by larger multiplicities of hadrons with a more democratic fragmentation, and therefore by a smaller value of PTD. The distribution of the quark-gluon discriminator in jets arising from the fragmentation of light quarks is studied in a control sample of $\gamma + $jet events, selected with stringent photon identification requirements and a b-jet veto, in which the residual small contamination from gluon jets is subtracted using the predictions from simulation; a good level of agreement between data and simulation is observed (Figure 7.3, left).

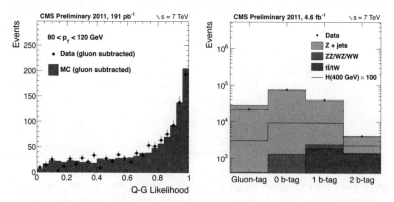

Figure 7.3. Left: distribution of the quark-gluon likelihood discriminant for light quark jets extracted from $\gamma + $jet events in the transverse momentum range 80–120 GeV/c^2. Right: multiplicity of $2\ell\,2$q events in the four flavour categories, in data and simulation; the open histograms is the expected distribution for a Higgs boson with mass 400 GeV/c^2 multiplied by a factor of 100.

The event multiplicities in the four flavour categories are shown in Figure 7.3, right. In general, a slight deficit in the gluon-tagged category is observed: this could arise from a different flavour composition of $Z + 2$jets events in data with respect to the simulation, or possibly from a difference in the quark-gluon distribution in data and simulations for gluon jets. Since the events in the gluon-tagged category are not used for the signal extraction, and normalization of the backgrounds is determined from data separately in each category, the discrepancy in the background categorization between data and simulation does not directly affect the result.

E_T^{miss} **significance.** In the two b-tag category, the contribution from dileptonic $t\bar{t}$ events to the background becomes sizable. In this analysis, the suppression is achieved using the missing transverse energy significance [52], defined as $S = 2 \ln \lambda$ where λ is the likelihood ratio of the two hypotheses of the true E_T^{miss} being equal to the measured value or to zero. This definition takes into account not only the measured value of E_T^{miss} and the total activity of the event $\sum E_T$, but also the detector resolution and the topology of the event. By construction, the distribution of S for all physics processes with no genuine E_T^{miss} is approximately the same, namely a χ^2 with two degrees of freedom, so that the same requirement $S < 10$ can be used for all Higgs boson mass hypotheses with a signal efficiency in the 97%–99% range.

7.1.2. Event selection in the low-mass region

For Higgs boson masses below $2m_Z$, one of the two Z bosons from the H decay is off-mass-shell. In this search, events are selected requiring the leptonically decaying Z to be off-shell ($m_{\ell\ell} < 80 \, \text{GeV}/c^2$), a requirement that also suppresses strongly the $Z + 2$jets background. The minimum p_T threshold on the two leptons are loosened to 20 and 10 GeV/c for the leading and subleading lepton, to compensate for the softer spectrum in this class of events.

The selection in the low-mass region is similar but simpler with respect to the one used at high masses. Since the discriminating power of the angular distribution is much reduced for events with an off-shell Z boson, the angular likelihood is not used is this mass range. Likewise, the gluon tagging is not used since performance of the algorithm is worse in this kinematic regime, characterized by jets of lower p_T and therefore less collimated. Since the kinematics of the two jets is also more uniform for the range of Higgs boson mass hypotheses tested, 130–165 GeV/c^2, the selection based on the E_T^{miss} significance in the events with two b-jets is replaced by the simpler requirement $E_T^{\text{miss}} < 50 \, \text{GeV}$.

7.1.3. Signal and background modelling

While in general a fair agreement is observed between the data and the expectations from simulations for the Z+2jets background, a more robust background estimation can be obtained by using directly the data in the sidebands of the m_{jj} distribution. For this purpose, the sidebands are defined as $60 < m_{jj} < 75\,\text{GeV}/c^2$ and $105 < m_{jj} < 130\,\text{GeV}/c^2$.

High-mass region. In the high-mass region, the normalization and shape of the background, $N_{\text{bkg}}(m_{ZZ})$, is obtained by extrapolating the measurement in the sidebands $N_{\text{sb}}(m_{ZZ})$ using the ratio of the predictions from simulation in the two regions,

$$N_{\text{bkg}}(m_{ZZ}) = N_{\text{sb}}(m_{ZZ}) \cdot \frac{N_{\text{sb}}^{\text{sim}}(m_{ZZ})}{N_{\text{bkg}}^{\text{sim}}(m_{ZZ})} = N_{\text{sb}}(m_{ZZ}) \cdot \alpha(m_{ZZ}). \quad (7.1)$$

The resonant diboson background is taken into account in the definition of $\alpha(m_{ZZ})$; its contribution to the overall background yield is anyway small, less than 5% in the categories with zero or one b-jet, and about 10% in the category with two b-jets. The extrapolation function α is predicted to to be in the range 0.75–1.2, reaching its maximum value around the ZZ threshold, and smoothly decreasing with increasing m_{ZZ}. Good consistency is observed between the predictions obtained with the two event generators MADGRAPH and SHERPA. A smooth background model is then obtained by fitting the prediction from equation (7.1) using an empirical parametrization.

A comparison between the observed m_{ZZ} distribution in data and the two predictions from this procedure and from the simulation is shown in Figure 7.4.

Low-mass region. At low mass, since the kinematic requirements are the same in all b-tagging categories, a single background shape is fitted to the combined events in the sidebands for all categories; this is necessary since the event yields are significantly smaller than in the high-mass range. The background normalization are then determined, separately in each category, directly from the events with m_{jj} in the signal region by using sidebands in the m_{ZZ} distribution. The $\alpha(m_{ZZ})$ correction is not needed anymore, since the overall normalisation is already determined, and the potential corrections to the shape alone are negligible in the narrow mass range considered.

The observed m_{ZZ} distribution and the corresponding background predictions from sidebands and simulation are shown in Figure 7.5.

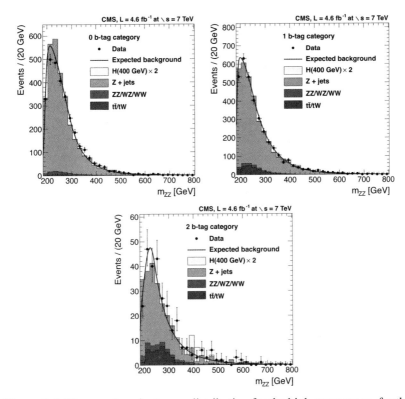

Figure 7.4. The m_{ZZ} invariant mass distribution for the high-mass range, for the three categories: 0 b-tag (left), 1 b-tag (middle), and 2 b-tag (right). Points with error bars show distributions of data and solid curved lines show the prediction of background from the sideband extrapolation procedure. Solid histograms depict the expectations from the different backgrounds and a hypothetical Higgs boson signal of mass 400 GeV/c^2 and twice the SM cross section.

Signal modelling. In both regions, the signal shape is obtained from a parametrization fitted to simulated events for several Higgs boson mass hypotheses, and then smoothly interpolated to intermediate Higgs boson masses. Theoretical uncertainties from the cross section and the lineshape for a heavy Higgs boson dominate the overall uncertainty on the signal yield. The largest uncertainties of experimental origin are from the knowledge of the jet energy scale and the b-tagging efficiencies and mistag rates.

7.2. H → ZZ → 2ℓ 2ν channel

The H → ZZ → 2ℓ ν decay mode has the second largest branching ratio among all ZZ modes considered at CMS, and the largest sensitivity for

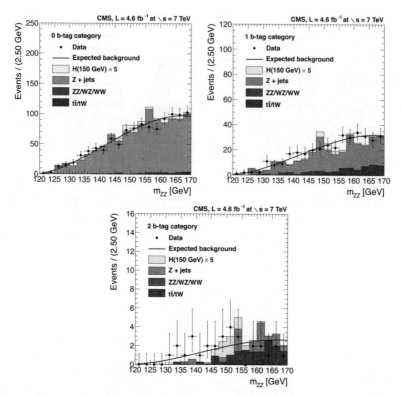

Figure 7.5. The m_{ZZ} invariant mass distribution for the low-mass range, for the three categories: 0 b-tag (left), 1 b-tag (middle), and 2 b-tag (right). Points with error bars show distributions of data and solid curved lines show the background shape extrapolated from the mjj sidebands, and scaled to the average normalization prediction from the m_{ZZ} sidebands. Solid histograms depict the expectations from the different backgrounds and a hypothetical Higgs boson signal of mass $150\,\mathrm{GeV}/c^2$ and five times the SM cross section.

heavy Higgs boson masses. When m_{H} is well above the $2m_Z$ threshold the two Z bosons are produced with a large transverse momentum, of order $m_{\mathrm{H}}/2\sqrt{2}$, resulting in a clean topology with a large missing transverse momentum from the invisible $Z \to \nu\nu$, back to back with a boosted $Z \to \ell^+\ell^-$.

The strategy of the search is simple: events are selected requiring a dilepton pair compatible with the decay of on-shell Z boson, and a large E_T^{miss}; the reducible background from $Z + \mathrm{jets}$ events is suppressed by requiring the $\vec{E}_T^{\mathrm{miss}}$ not to be aligned with a hadronic jet, and the $t\bar{t}$ and WZ backgrounds are controlled by vetoing the presence of b-jets and additional leptons. Signal extraction is performed by counting events in a window of a transverse mass variable built out of the lepton momenta and the E_T^{miss}, after a categorization based on the number of jets and their topology.

7.2.1. Event selection

Events are selected online using double electron, single muon and double muon triggers; the trigger efficiency for events with both leptons in the p_T and η acceptance of the offline analysis is above 95% and 99% for dimuon and dielectron events respectively, as measured from $Z \to \ell^+\ell^-$ events in data.

Leptons. Offline, events are selected requiring two leptons with the same flavour and with opposite electrical charges, with $p_T > 20\,\text{GeV}/c^2$. Muons are accepted in the full $|\eta| < 2.4$ region, while electrons are accepted for $|\eta| < 2.5$ but excluding the transition region between barrel and endcaps ($1.444 < |\eta| < 1.566$). Tight identification and impact parameter requirements are applied to both leptons, complemented by isolation criteria defined using particle candidates reconstructed from the particle flow algorithm in a η-ϕ cone of radius 0.4 centered on the lepton; contaminations from pile-up in the isolation cone are subtracted statistically using a median per-event energy density ρ (Section 2.4.4). The selection efficiency for muons and electrons are measured as function of p_T and η from $Z \to \ell^+\ell^-$ events, typical values ranging between 90% and 97% for muons, and between 70% and 90% for electrons.

In addition to the p_T thresholds on the individual leptons, the p_T of the dilepton system is also required to be larger than $55\,\text{GeV}/c^2$, and the dilepton invariant mass to be within $15\,\text{GeV}/c^2$ from the nominal Z boson mass.

Missing energy. The missing transverse energy is reconstructed from the particle-flow algorithm, and rejection criteria are applied against anomalous calorimetric noise and beam-halo events. For the lightest Higgs boson mass hypotheses considered in the analysis, $m_\text{H} = 250\,\text{GeV}/c^2$, the E_T^{miss} is required to be larger than $70\,\text{GeV}$; the requirement is increased linearly with m_H up to $80\,\text{GeV}$ for $m_\text{H} = 300\,\text{GeV}/c^2$, and kept constant afterwards.

To suppress backgrounds from $Z + $ jets production where the momentum unbalance arises from the underestimation of the energy of a hadronic jet, the azimuthal angle between the $\vec{E}_T^{\text{miss}}$ vector and the closest jet with $p_T > 30\,\text{GeV}/c^2$ is required to be larger than 0.5 radiants; if no jets with $p_T > 30\,\text{GeV}/c^2$ are found, the requirement is applied to the closest jet with $p_T > 15\,\text{GeV}/c^2$. In the 2012 data, events are vetoed also if the missing transverse energy is aligned with one of the leptons,

$\Delta\phi(\ell, \vec{E}_T^{\mathrm{miss}}) < 0.2$, to reject backgrounds where the unbalance is created by a large mismeasurement of the lepton energy.

A transverse mass variable m_T is defined from the dilepton transverse momentum $p_T^{\ell\ell}$, dilepton mass $m_{\ell\ell}$ and the $\vec{E}_T^{\mathrm{miss}}$ vector as follows:

$$m_T^2 = \left(\sqrt{(p_T^{\ell\ell})^2 + (m_{\ell\ell})^2} + \sqrt{(E_T^{\mathrm{miss}})^2 + (m_{\ell\ell})^2} \right)^2$$
$$- \left(\vec{p}_T^{\,\ell\ell} + \vec{E}_T^{\mathrm{miss}} \right)^2. \tag{7.2}$$

Events are selected requiring the m_T to be in a range dependent on the Higgs boson mass hypothesis.

Other requirements.   Rejection of $t\bar{t}$ and tW background events is done by vetoing events containing a b-tagged jet with $p_T > 30\,\mathrm{GeV}/c^2$ and $|\eta| < 2.4$; similarly to the other Higgs boson searches, the track counting high efficiency b-tagger is used for this purpose. This veto is complemented with one triggered by soft muons arising from decays of B and D hadrons, as in the $H \to WW \to 2\ell 2\nu$ analysis. The combination of the two requirements provides a suppression factor of about six for this class of backgrounds.

A veto on additional electrons and muons with $p_T > 10$ satisfying the nominal lepton selection is applied to control backgrounds from from $WZ \to 3\ell\,\nu$ events, and to a lesser extent $ZZ \to 4\ell$ with E_T^{miss} arising from an undetected lepton.

Event categorization.   First, a VBF event category is defined by requiring two or more hadronic jets with $p_T > 30\,\mathrm{GeV}/c$, with a $|\Delta\eta| > 4$ requirement between the two closest jets; the invariant mass of these two tagging jets must exceed $500\,\mathrm{GeV}/c^2$, and the leptons from the Z decay are required to be in the η region between the two jets. Events not selected in the VBF category are then subdivided according to the number of hadronic jets with $p_T > 30\,\mathrm{GeV}/c$ into three categories with zero, one, and two or more jets.

The expected signal composition in the VBF category is about 75% from VBF production and 25% from the gluon fusion production, for a Higgs boson mass hypothesis of $400\,\mathrm{GeV}/c^2$. The signal yields in the other categories are dominated by gluon fusion; the contamination from VBF production is 8–10% in the final states with at least one jet, and about 2% in the final state with no jets.

7.2.2. Background estimation

The backgrounds to the $H \to ZZ \to 2\ell 2\nu$ search can be classified in three categories: Z+jets with instrumental E_T^{miss}; non-resonant dileptons from WW, t$\bar{\text{t}}$ and tW; and resonant diboson production, including the irreducible ZZ $\rightarrow$ 2ℓ2 ν.

Z+jets background estimation. This background is estimated from data relying on the more abundant γ + jets events, for which the momentum imbalance is purely instrumental as for Z + jets. For this estimation, events are selected requiring a well identified and isolated photon satisfying the same p_T requirement applied to the dilepton system (p_T > 55 GeV/c). A jet of p_T > 15 GeV/c is also required, to suppress the W/Z + γ events with genuine missing energy from W or Z decaying to neutrinos, and similarly W $\rightarrow$ ℓ ν events with a single jet misidentified as a photon.

To reproduce the kinematic of the background, the γ + jets events weighted as function of the photon p_T to match the transverse momentum spectrum of the Z bosons in Z + jets events selected inclusively in data with no E_T^{miss} requirements; the procedure is done separately in categories defined by the number of jets with p_T > 30 GeV/c, since the E_T^{miss} performance is dependent on the overall hadronic activity. A further weighting is applied as function of the number of primary vertices, to correct for the residual pile-up differences.

The predicted E_T^{miss} distribution for dilepton events events obtained using γ + jets events is found to be in excellent agreement with what observed in the data (Figure 7.6). The corresponding m_T distribution is obtained by applying the definition in equation (7.2), using the photon p_T in place of $p_T^{\ell\ell}$, and by randomly sampling a $m_{\ell\ell}$ value from the inclusive dilepton mass spectrum of Z $\rightarrow$ $\ell^+\ell^-$ events in data.

The residual contaminations of events with real E_T^{miss} in the γ + jets events, e.g. from W + γ or Z + γ production followed by W $\rightarrow$ ℓ ν or Z $\rightarrow$ ν ν decays, can result in an overestimation of this background. This effect is corrected empirically by reducing the predicted background normalization by a factor two and assigning a 100% systematical uncertainty to it.

Non-resonant background estimation. Backgrounds where the two leptons do not arise from the decay of a Z boson are predicted using e μ events in data satisfying the full selection.

The extrapolation factor to the μ μ and ee final states is obtained from the corresponding ratios of $\ell\ell$ to e μ events in sidebands of the dilepton mass distribution, $m_{\ell\ell}$ $\in$ [40, 70] $\cup$ [110, 200] GeV/c^2, with the requirement E_T^{miss} > 70 GeV; to further enrich the purity of t$\bar{\text{t}}$ events in the sidebands, events are required to contain a b-tagged jets.

Resonant diboson backgrounds. The contributions from resonant diboson production is estimated using simulated events, generated us-

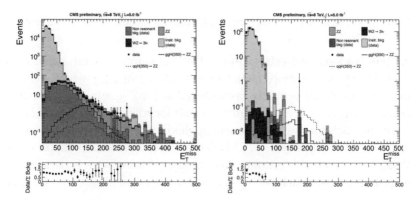

Figure 7.6. Observed E_T^{miss} distributions in 8 TeV data, and predictions for the Z + jets events, determined from γ + jets events in data, and the other backgrounds, predicted from simulations. The expected distribution for a Higgs boson signal of mass $m_{\mathrm{H}} = 350\,GeV$ is also shown. The plot on the right is for events in the VBF category, the one on the left for the remaining events, inclusively on the number of hadronic jets.

ing MADGRAPH interfaced with PYTHIA for showering and hadronization. In the case of WZ events, the contribution is normalized to the NLO predictions for the total cross section. For ZZ events, a dynamical k-factor is applied as a function of the transverse mass m_T, determined from MCFM[91], and the cross sections is further scaled by to account also for the gg $\to$ ZZ production mechanism (a 6% correction).

7.2.3. Signal extraction and systematical uncertainties

Signal extraction is performed by counting events in a m_T window dependent on the Higgs boson mass hypothesis of choice, after a E_T^{miss} requirement also dependent on the signal mass hypothesis. The ee and $\mu\,\mu$ final states are analyzed separately. In the VBF category, due to the very low expected signal and background yields, the same E_T^{miss} requirement is used for all Higgs boson mass hypotheses, and no m_T requirement is applied; the results at each Higgs boson mass hypohesis therefore differ only because of the dependency of the expected signal yield on m_{H}. For illustration, the m_T distributions in two of the final states are shown in Figure 7.7.

The predictions for the signal model are taken from simulated events, generated using POWHEG, weighed as function of the Higgs boson p_T to reproduce the NNLL+NLO predictions from HQT, and normalized to the most accurate predictions for the total production cross section from reference [10]. The simulated events are further re-weighted to match the luminosity spectrum of the data.

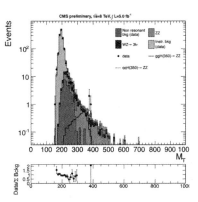

Figure 7.7. Observed m_T distributions in 8 TeV data, and corresponding signal and background predictions. The left plot is for the $\mu\,\mu$ final state with no hadronic jets, while the one on the right is for the ee final state in the VBF category.

While the $2\ell\,2\,\nu$ final state could potentially receive contributions also from the H $\rightarrow$ WW $\rightarrow$ $2\ell\,2\,\nu$ decays, this signal is not included in the search, as it would anyway be subtracted together with the non-resonant backgrounds estimated from e μ events.

Systematical uncertainties on the Higgs boson production cross section from the neglected higher orders of the perturbative expansion and from the knowledge of the parton density functions are taken from reference [10], while the ones arising from the modelling of the lineshape for heavy Higgs boson signals are evaluated according to the prescriptions in reference [83]. A systematical uncertainty on the shape of the m_T distribution for signal is obtained by weighting the events to different Higgs boson p_T spectra predicted from HQT for different values of the factorization and renormalization scales.

For the signal and the diboson background estimated from simulations, additional systematical uncertainties are assigned from the knowledge of lepton reconstruction and selection efficiencies, lepton and jet energy scales, b-tagging efficiency, and normalisation of the integrated luminosity.

7.3. H $\rightarrow$ WW $\rightarrow$ $\ell\nu q\bar{q}$ channel

For a heavy Higgs boson, the H $\rightarrow$ WW $\rightarrow$ $\ell\,\nu q\bar{q}$ final state has the largest branching fraction among all modes analyzed at the LHC. The lepton in the initial state provides a handle to select these events at trigger level, and the presence of a single neutrino allows the full kinematic to be reconstructed from the missing transverse energy, up to a two-fold ambiguity from the longitudinal component of the neutrino momenta, above the threshold for the production of two on-shell W bosons.

The analysis is performed by searching for an excess in the four-body invariant mass reconstructed from the momenta of two jets, the lepton and the neutrino, after a selection on kinematic and topological variables, including the angular spin correlation between decay products. The main challenge to this search is given by the W + 2jets background, whose production cross section is about two orders of magnitude larger than the signal one. Due to the rapidly falling four-body mass spectrum of the background, a good sensitivity is achieved for heavy Higgs boson mass hypotheses, $m_H \sim$ 300–450 GeV/c^2; conversely, the mode is not viable, and not pursued, for Higgs boson masses below the WW threshold.

7.3.1. Event reconstruction and selection

Events are selected by the requirement of a single isolated and well identified electron or muon, missing transverse energy from the undetected neutrino, and the presence of two or three central hadronic jets.

Leptons. In the analysis of the 2011 and 2012 datasets have not been performed simultaneously, for historical reasons different lepton identification and isolation criteria have been used, although tuned for similar efficiencies. For muons, the tight selection described in Section 2.4.3 is used, except for a minor re-tuning of the criteria to account for the differences in the track reconstruction algorithms between the two datasets, For electrons, the 2011 identification is based on simple tight requirements on the main electron identification variables, while in 2012 data the new multivariate electron identification (Section 2.5.4) has been used.

Isolation criteria are applied on the basis of the energy flow in a cone around the leptons: in the analysis of the 2011 dataset, tracks and calorimetric deposits are used, while for the 2012 analysis particle flow objects are used.

To suppress backgrounds from Drell-Yan, $t\bar{t}$ and diboson production, a veto is applied on events with additional leptons besides the one from the candidate W $\rightarrow \ell\nu$ decay. A looser lepton identification and isolation criteria is used for this purpose.

Jets. As common throughout the CMS analyses, jets are reconstructed with the anti-k_T algorithm with distance parameter $R = 0.5$ from the output of the particle flow reconstruction. Jet energy corrections for pile-up based on the median per-event energy density are applied, combined with traditional calibrations for the detector response (Section 2.6.1). Jet identification criteria to suppress signal from anomalous detector noise and from pile-up are also used.

Jets are considered only within the acceptance of the inner tracker, $|\eta| < 2.4$, and are required to have $p_T > 30$ GeV. Events are selected requiring the presence of at least two hadronic jets, and the leading two are used to define the candidate $W \to q\bar{q}$ decay. Studies on simulations reveal that this assignment is correct in 68–88% of the cases for Higgs boson masses in the 200–600 GeV/c^2 range for events with only two central jets.

As final state radiation can result in additional jets, events with three jets are also accepted by the selection, but events with more than three are vetoed to suppress semi-leptonic $t\bar{t}$ decays. In three-jet events, the correctness of the assignment of the two leading jets to the W decay is lower, ranging from 26% at $m_{\rm H} = 200$ GeV/c^2 to 84% at 600 GeV/c^2.

Missing transverse energy.   In addition to the jets and leptons, a requirement on the missing transverse energy is applied: $E_T^{\rm miss} > 25$ GeV for final states with muons, and $E_T^{\rm miss} > 30$ GeV for events with electrons, the difference being motivated by the larger QCD multi-jet background in the electron case. The transverse mass of the W boson reconstructed from the lepton and the missing energy is required to be larger than 30 GeV/c^2, to suppress other non-W backgrounds.

Kinematic fit.   The invariant mass of the dijet pair from the hadronic W decay is required to be in the 65–95 GeV/c^2 range, corresponding to an efficiency of about 80% for signal events.

In order to improve the resolution on the reconstructed four-body mass, a kinematic fit is performed on the momenta of the lepton and the two jets and to the missing transverse energy vector, constraining the two W bosons to be on shell; the natural width of the W is taken into account in the fit procedure. The fit improves significantly the discrimination between a Higgs boson signal and the continuum W + jets and WW background for masses close to the kinematic threshold (Figure 7.8)

7.3.2. Signal extraction

A multivariate likelihood discriminator is used to discriminate between the Higgs boson signal and the main W+jets background. The discriminator relies on the p_T and rapidity of the WW system, on five angles parameterizing the decay kinematic, and on the lepton charge to benefit from the asymmetry in W + jets production. In the 2011 analysis, the discriminator to separate quark jets from gluon jets used in the H $\to$ ZZ $\to$ $2\ell 2$q is also included in the multivariate analysis. The discriminator is trained on simulated events separately in each final state for different Higgs boson mass hypotheses, and events are selected if the output value of the discriminator is above a threshold obtained optimizing for the best exclusion sensitivity.

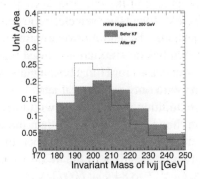
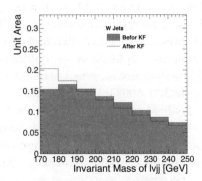

Figure 7.8. Four-body invariant mass distribution for a Higgs boson signal of mass 200 GeV/c^2 (left) and for the continuum W + jets background (right). The mass before the kinematic fit is shown as a solid histogram, while the mass after it as an outlined histogram. The distribution for the WW background is very similar to that of W + jets.

The signal extraction is performed from a binned analysis of the four-body invariant mass distribution (Figure 7.9, right). Templates for the dominant W + jets background are determined from data, using a linear combination of the high and low dijet mass sidebands for the W $\rightarrow$ q$\bar{\text{q}}$ decay, with the coefficient determined from studies on simulated events. All other templates are determined from simulations, except for the reducible QCD multi-jet background, extracted from data by inverting the lepton isolation criteria and relaxing the identification and E_T^{miss} requirements.

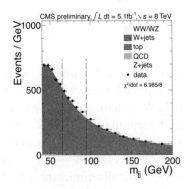
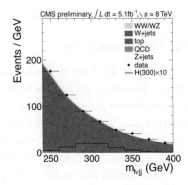

Figure 7.9. Invariant mass distributions in the μ + 2jets final state in 2012 data, after the likelihood selection for a Higgs boson mass hypothesis of 300 GeV/c^2: dijet mass used to determine the W + jets yield (left) and four-body mass used for signal extraction (right).

The normalization of the W + jets background is extracted from a fit to the dijet invariant mass in data (Figure 7.9, left), excluding the signal region. In this fit, the W + jets shape is modelled using simulated events for Higgs boson mass hypotheses below $200\,\text{GeV}/c^2$, and as an empirical function fitted to simulated events for higher masses where the number of simulated events passing the likelihood selection becomes too limited. The contribution from the reducible multi-jet background in this fit is constrained to what measured from data by fitting the W transverse mass distribution, within a large uncertainty of 50/100% for electrons and muons respectively, and its shape determined from data as for the four-body mass shape.

The diboson, top and Z + jets background are constrained to their theoretical predictions within the respective uncertainties, both in the dijet mass sideband fit and in the signal extraction from the four-body invariant mass distribution.

The dominant sources of uncertainty in the selection are from the knowledge of the shape of the W + jets background, the theoretical uncertainties on the signal production cross section, lineshape and jet multiplicity distribution, and on the efficiency of the likelihood selection on signal events.

As a cross-check the analysis is repeated leaving the W + jets background contribution unconstrained in the four-body mass shape, yielding compatible results but with larger uncertainties, as the broadness of the reconstructed lineshape of a Higgs boson signal does not allow for a very good discrimination between signal and background.

The analysis has been performed separately on the 2011 and 2012 dataset, and on the combination of the two, probing for Higgs boson mass hypotheses in the 180–$600\,\text{GeV}/c^2$ range. No significant deviations are observed with respect to the expectations from the background-only hypothesis.

Chapter 8
Statistical analysis for Higgs boson searches

Statistical inference is used to translate the outcome of the searches for a Higgs boson into statements about evidence or exclusion of a signal.

The first step of the procedure is to clearly define how signals, backgrounds and systematical uncertainties are modelled, so that this information can be encoded in mathematical form into a likelihood function expressing the probability, or probability density, of an experimental outcome as function of all the unknown parameters in the model. Afterwards, statistical methods are used to convert this information into exclusion limits and statistical significances.

The statistical methodology for the analysis of the Higgs boson searches at the LHC has been developed by the ATLAS and CMS collaborations in the context of the LHC Higgs Combination Group [76].

After the combined analysis of 2011 and 2012 resulted in the observation of a new particle, the focus of the analysis has been extended to the measurement of the properties of this particle to assess if it is compatible with the predictions for a SM Higgs boson. The additional statistical methodology used for this task, not included in the LHC Higgs Combination Group report, is described in the last section of this chapter.

8.1. Signal model

The standard model is a strongly constrained theory, in which the mass of the Higgs boson is the only unknown parameter[1]: once that is specified, the Higgs boson production cross sections, branching fractions and kinematics are fully defined. However, it is customary to consider the slightly more general scenario of a SM-like Higgs boson in which the cross sections for all production modes are scaled by a single factor

---

[1] Indirect information on the Higgs boson mass exits from the electroweak precision fits, but is not used in the direct searches. Also, there are other unknown parameters in the neutrino sector, but they are irrelevant in this context.

$\mu = \sigma/\sigma_{SM}$ compared to the SM expectations, preserving the branching fractions. From the phenomenological point of view, this scenario can accommodate for modifications from new physics beyond the standard model affecting the production rates or creating new decay channels and therefore reducing the branching fractions in the modes searched for. Practically, this presentation is also useful because the exclusion limits on μ from a search convey information about its sensitivity also when it is it far from the one needed to exclude a SM Higgs signal; these limits also obey the same intuitive scaling rules as limits on cross sections, allowing to perform simple extrapolations of the expected performance for different integrated luminosities, signal efficiencies or background levels.

When considering analyses that are sensitive to a single production and decay mode, the limits on σ/σ_{SM} can also be directly presented as model-independent limits on the production rate for a given final state. When multiple modes contribute, assumptions from the model are used to combine the results together, and this interpretation is no longer possible or useful.

The search for a SM Higgs boson at LHC is performed at discrete values of the Higgs boson mass, chosen at steps narrower than the observable width of a possible signal. The natural width of a light SM Higgs boson is small compared to the experimental resolution; therefore, in this case the choices of Higgs mass points used in the search is determined by the channels with the best mass resolution, *i.e.* H $\rightarrow$ $\gamma\gamma$ and H $\rightarrow$ ZZ $\rightarrow$ 4 μ, for which $\Delta m_H/m_H \sim 1\%$. For masses above about 250 GeV/c^2, the natural width of the signal becomes larger than the experimental resolution, and drives the choice of steps used. The mass points used in the CMS search are listed in Table 8.1.

mass range (GeV/c^2)	step size (GeV/c^2)	number of points
110–150	0.5	80
150–160	1	10
160–290	2	65
290–350	5	12
350–400	10	5
400–600	20	11

Table 8.1. Higgs boson mass hypotheses probed in the CMS searches.

8.2. Systematical uncertainties

Systematical uncertainties, in this context, are all the quantified sources of uncertainty in the signal and background expectations to which the observed data are compared. This includes theoretical uncertainties on the expected cross sections and acceptances for signal and background processes, experimental uncertainties arising from modelling of the detector response (event reconstruction and selection efficiencies, energy scale and resolution), and statistical uncertainties associated with either ancillary measurements of backgrounds in control regions or selection efficiencies obtained using simulated events. Systematic uncertainties can affect event yields, expected shapes of the distributions, or both.

When combining the results of different analyses together, it is important to account correctly for the correlation of the effects that systematical uncertainties have across the analyses. This is done by factorizing the uncertainties into independent sources, and assuming that the effect of a single source is correlated across all channels where it is relevant. A guiding principle when arbitrating if two uncertainties are to be taken as correlated or uncorrelated is that assuming a positive correlation is more conservative than assuming no correlation, and no correlation is more conservative than an anti-correlation[2].

In the combined analysis of the data from the 2011 and 2012 runs, it is also important to assess the correlation between the systematical uncertainties for the analyses of the two periods. The systematical uncertainties arising from theoretical predictions or from assumptions used in estimating the backgrounds (*e.g.* the universality of the lepton misidentification probabilities) are taken to be fully correlated across the two periods, while no correlation is assumed for uncertainties of statistical origin or when very different methods are used.

Several systematic uncertainties are correlated across multiple search channels, and in particular:

- The uncertainty on the absolute scale of the integrated luminosity is correlated across all channels. However, no correlation is assumed between the 2011 and 2012 normalizations since the common uncertainty from the beam current normalization is small compared to the other systematical uncertainties in the two methods used[3].
- Experimental uncertainties on selection and trigger efficiencies for the same kind of physics object are assumed to be correlated, even if the

---

[2] This is true if the data in the signal region does not constrain significantly the uncertainty being considered; this condition is easily satisfied *e.g.* by all uncertainties related to the signal model, unless an evident signal is present in the data.

[3] The 2011 luminosity scale is normalized from an offline calibration based on the rate of clusters in the pixel subdetector, while for the 2012 luminosity at the moment only the less accurate measurement from the forward hadronic calorimeter is available.

selections used are sometimes different. It should be noted the correlation does not imply that the size of the uncertainty is the same.

- Energy scale and resolution uncertainties are likewise assumed to be correlated across objects of the same kind. No correlation is assumed between photon and electron energy scales, since the uncertainties are driven by the higher level calibrations, for which significantly different methods are used for the two, and the kinematic regime is also different[4].

- Theoretical uncertainties on the production cross sections for each process arising from the unknown higher orders terms in the perturbative series are totally correlated across the analyses. The uncertainties on different processes, *e.g.* gluon fusion and VBF Higgs boson production, are assumed to be independent; for this purposes, W and Z bosons are considered to be the same particle, since the computations for processes with identical final state except for that are usually closely related, *e.g.* W + H and Z + H associated production.

- Theoretical uncertainties from the knowledge of the parton density functions are assumed to be correlated across processes that have the same partonic initial state (qq/q$\bar{\text{q}}$, gg, qg), and independent otherwise. The validity of this approximation has been checked against the full covariance matrix of PDF uncertainties for different processes computed by the CTEQ collaboration [101], as described in reference [76].

- Theoretical uncertainties on the lineshape for a heavy Higgs boson are totally correlated across all production and decay modes.

8.3. Likelihood functions

The likelihood function of an analysis or of the full combination, for a fixed value of the Higgs boson mass, is defined in terms of the observed data, the signal strength modifier $\mu = \sigma/\sigma_{SM}$, and the signal and background models. In general, it is not possible to describe the CMS searches at all masses using a single likelihood function depending also on the hypothesized value of the mass, since several analyses have event selection requirements dependent on the mass and therefore there is no single observed dataset for all masses.

When constructing the likelihood function, each independent source of systematical uncertainty is assigned a nuisance parameter θ_i, and the

---

[4] The sensitivity in the diphoton search is driven by high momentum and high quality photons, while the four-lepton analysis relies on low p_T electrons with loose requirements: the extrapolation of the reference energy calibrations from the Z $\to$ e$^+$e$^-$ standard candle are therefore significantly different for the two.

expected signal and background models are described in terms of those parameters. Usually, a systematical uncertainty reflects the possible deviation of a quantity from the input value obtained from a separate measurement: in these cases, the information from the other measurement is represented through a variable $\tilde{\theta}_i$, representing the outcome of the other measurement, and a probability density function $p_i(\tilde{\theta}_i|\theta_i)$ describing the probability to measure a value $\tilde{\theta}_i$ given the true value θ_i. When instead all the knowledge about a parameter of the model comes from the data in the signal region, as is the case for the background normalization and shape in H $\rightarrow$ $\gamma\gamma$, the corresponding nuisance parameter is left unconstrained, with no associated $\tilde{\theta}_i$ and $p_i(\tilde{\theta}_i|\theta_i)$.

The likelihood function is in general written as the product of likelihoods of the data in each channel times the product of the probability densities for the measurements associated to the nuisance parameters,

$$L(\text{data}, \tilde{\theta} \mid \mu, \theta) = \prod_{n=1}^{N_C} L_n(\text{data}_n \mid \mu, \theta) \times \prod_{i=1}^{N_\theta} p_i(\tilde{\theta}_i|\theta_i) . \quad (8.1)$$

Binned data. The likelihood function for channels that use binned data, including counting experiments, is the product of Poisson terms in each bin k, defined by the observed events n_k and expected signal and background yields $s_k(\theta), b_k(\theta)$:

$$L(\text{data} \mid \mu, \theta) = \prod_{k=1}^{N_B} \text{Poisson}(n_k \mid \mu \cdot s_k(\theta) + b_k(\theta)) . \quad (8.2)$$

Unbinned data. In the analysis of some channels, e.g. H $\rightarrow$ ZZ $\rightarrow$ 4ℓ, events are used in an unbinned way, in order to exploit all the information about the individual value of a quantity x, e.g. the invariant mass of the system. The likelihood function for these channels is expressed in terms of the observed events n, the expected yields $s(\theta)$ and $b(\theta)$, and also the probability density functions of the variable x for signal and background events $f_S(x|\theta)$ and $f_B(x|\theta)$: the overall probability density function is defined from the normalized distributions of signals and backgrounds,

$$f(x \mid \mu, \theta) = \frac{\mu \cdot s(\theta)}{\mu \cdot s(\theta) + b(\theta)} f_S(x \mid \theta) + \frac{b(\theta)}{\mu \cdot s(\theta) + b(\theta)} f_B(x \mid \theta), \quad (8.3)$$

and the likelihood is given by a Poisson term times a product of probability densities on all events e:

$$L(\text{data} \mid \mu, \theta) = \text{Poisson}(n \mid \mu \cdot s(\theta) + b(\theta)) \times \prod_{e=1}^{N_E} f(x_e \mid \mu, \theta) . \quad (8.4)$$

It can be shown that the definitions in equations (8.2) and (8.4) are identical in the limit of infinitely fine bin sizes (or piece-wise constant densities f_S, f_B), up to irrelevant multiplicative terms that depend only on the observed data and not on any parameter μ, θ.

Systematical uncertainties.   Two classes of probability density functions have been used to model the external measurements that constrain the nuisance parameters, depending on the origin of the uncertainty. For nuisances representing the statistical uncertainties on the number of events in a control region or simulated sample, $\tilde{\theta}_i$ is taken to be the observed event count and θ_i the expected yield, so that $p_i(\tilde{\theta}_i | \theta_i)$ is a Poisson probability.

In all other cases, Gaussian probability density functions are used. A choice of parametrization is made so that θ_i and $\tilde{\theta}_i$ are dimensionless, the variance of the Gaussian is one, and the observed value $\tilde{\theta}_i$ is zero. When parameterizing the effect of a variation of θ_i into a change of a positive definite physical observable $\mathcal{O}$, e.g. an event yield, an exponential parametrization $\mathcal{O} = \mathcal{O}_0 \cdot \kappa^{\theta_i}$ is used, where κ is a dimensionless constant quantifying the size of the uncertainty ($\kappa \simeq 1 + \Delta\mathcal{O}/\mathcal{O}$); the resulting probability density for $\mathcal{O}$ is a log-normal.

8.4. Statistical inference

Two complementary paradigms exist about how statistical inference is performed, which stem from two different interpretations of the concept of probability.

- In the frequentist paradigm, the probability is seen as the asymptotic value for the frequency of the outcomes of a large number of identical experiments, and therefore frequentist inference is used to make statements about the probability of an experimental outcome for a given model.

  The exclusion a SM Higgs boson signal at 95% confidence level (CL) is to be interpreted in this context as the statement that the probability of obtaining an outcome as background-like as the observed one if the signal hypothesis were to be true is 5% or less; in the remaining 95% or more of the experimental outcomes, the presence of the signal would have been more evident than what observed in the data.

- In the Bayesian paradigm, the probability is interpreted as a subjective degree of belief in the validity of a theory: in this context, statements are made about the probability of a given model to be true given the observed experimental outcome and the subjective prior assumption about that probability before the experiment was performed.

The exclusion of the SM Higgs boson cross section at 95% credible level (also CL), assuming a flat prior probability on the cross section, has the following meaning: if the experimental outcome and the assumed prior are combined by means of the Bayes theorem to produce a posterior probability density for the Higgs boson cross section, 95% of the probability density is for a cross section values smaller than the SM one.

In the interpretation of the outcomes of the searches for a SM Higgs boson at LHC, it has been decided to follow preferentially the frequentist paradigm, but a Bayesian interpretation of the same outcomes is also provided. The agreement between the results obtained with the two paradigms is a sign of the robustness of the interpretation of the data.

In the frequentist approach, limits and significances for a given Higgs boson mass hypothesis are computed by means of a test statistics, a function which summarizes in a single number the information on the observed data, expected signal, expected background, and all uncertainties associated with these expectations; the test statistics allows one to rank all possible experimental observations according to whether they are more consistent with the background-only or signal+background hypotheses. In order to infer the presence or absence of a signal in the data, the observed value of the test statistic is compared with the distributions expected under the background-only and signal+background hypotheses.

The expected distributions of the test statistics are obtained by generating an ensemble of pseudo-datasets from the probability densities used to define the likelihood in eq. (8.2), (8.3); in each pseudo-dataset, new values of the measurements $\tilde{\theta}_i$ are also generated, using the probability densities $p_i(\tilde{\theta}_i|\theta_i)$. The values of the nuisance parameters θ used for generating the pseudo-datasets and $\tilde{\theta}$ are obtained by maximising the likelihood for the observed data and $\tilde{\theta}$ under the background-only or signal+background hypothesis.

This methodology is characterized by a frequentist treatment of the systematical uncertainties, in which the measurements $\tilde{\theta}$ are sampled, and the parameters θ are kept constant. In this respect, it differs from the conventional hybrid frequentist-Bayesian procedure [102]: in the hybrid approach, each pseudo-dataset is generated using a different value of the nuisance parameters θ sampled from a probability density $\tilde{p}_i(\theta_i|\tilde{\theta}_i)$, obtained inverting $p_i(\tilde{\theta}_i|\theta_i)$ by means of Bayes theorem, while $\tilde{\theta}$ are kept constant. However, despite this important difference, in practice the two approaches yield very similar results.

8.5. Quantifying a local excess

In order to quantify the statistical significance of an excess over the background-only expectation, a test statistic q_0 is defined using a likelihood ratio with the background-only hypothesis in the numerator, and an unconstrained signal-plus-background hypothesis in the denominator:

$$q_0 = -2\ln\frac{L(\text{data}, \tilde{\theta} \mid 0, \hat{\theta}_0)}{L(\text{data}, \tilde{\theta} \mid \hat{\mu}, \hat{\theta})} \quad \text{with a constraint } \hat{\mu} \geq 0, \quad (8.5)$$

where $\hat{\theta}_0, \hat{\theta}$, and $\hat{\mu}$ are the values of the parameters θ and μ that maximise the likelihoods in the numerator and denominator, and the subscript in $\hat{\theta}_0$ indicates that the maximization in the numerator is done under the background-only hypothesis ($\mu = 0$). The profiling of the nuisance parameters, i.e. maximization of the likelihood as function of θ, improves the sensitivity by exploiting the information from the data in the signal region to improve the knowledge on θ beyond what was available from the external measurements $\tilde{\theta}$ alone.

The definition (8.5) departs from the one used in the analysis of the Higgs boson searches at LEP and Tevatron in particular in the use of $\hat{\mu}$ at the denominator, instead of a fixed $\mu = 1$ corresponding to the SM signal expectation. This choice is motivated by the existence of analytical results for the expected distribution of q_0 under the background-only hypothesis in the asymptotic limit of a large number of events, as explained below.

With this definition, a signal-like excess corresponds to a positive value of q_0, while a deficit with respect to the background-only expectation yields a q_0 value of zero. An excess can then be quantified in terms of the p-value p_0, which is the probability to obtain a value of q_0 at least as large as the one observed in data, q_0^{obs}, under the background-only hypothesis.

$$p_0 = P\left(q_0 \geq q_0^{\text{obs}}\right). \quad (8.6)$$

An example distribution of the test statistic q_0 for one of the Higgs boson mass hypothesis tested is shown in Figure 8.1.

The significance Z is obtained from the p-value via the Gaussian one-sided tail integral:

$$p_0 = \int_Z^\infty \frac{1}{\sqrt{2\pi}} \exp(-x^2/2) \, dx. \quad (8.7)$$

The test statistic q_0 has one degree of freedom (μ) and, in the limit of a large number of events, its distribution under the background-only hypothesis converges to a half of the χ^2 distribution for one degree of freedom

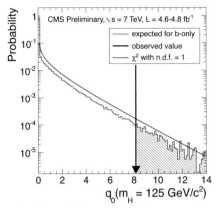

Figure 8.1. Observed value of the test statistics q_0 and expected distribution under the background-only hypothesis, for the combination of the CMS Higgs boson searches on 2011 data at $m_H = 125\,\text{GeV}/c^2$. The local p-value is computed from the integral of the expected distribution of q_0 above the observed value (hatched area), and corresponds to a local significance of about $3.0\,\sigma$. The χ^2 distribution with one degree of freedom is also shown; the significance estimated from the observed value of the test statistics using the χ^2 distribution is about $2.8\,\sigma$, in fair agreement with the other result.

plus $0.5 \cdot \delta(q_0)$, where $\delta(q_0)$ stands for the delta-function [103]. This asymptotic property allows the significance to be evaluated directly from the observed test statistic q_0^{obs} as

$$Z = \sqrt{q_0^{\text{obs}}}. \tag{8.8}$$

The observed p-value obtained with this definition is independent of the normalization of the expected signal, since the likelihood function is evaluated only for $\mu = 0$ and $\mu = \hat{\mu}$. In the context of the SM Higgs boson searches, however, the expected signal rate is known, so it is useful to quantify the magnitude of an observed excess in terms of the signal expectations. This is expressed in terms of the value $\hat{\mu}$ that best describes the data obtained from a maximum likelihood fit; in order to quantify also the size of the deficits, the constraint $\mu \geq 0$ is relaxed. An approximate 68% confidence level interval for μ around the best fit value $\hat{\mu}$ can be determined from the condition

$$-2\ln \frac{L(\text{data}, \tilde{\theta} \mid \mu, \hat{\theta}_\mu)}{L(\text{data}, \tilde{\theta} \mid \hat{\mu}, \hat{\theta})} \leq 1 \tag{8.9}$$

as in the asymptotic regime the left hand side of the inequality is distributed as a χ^2 with one degree of freedom.

8.6. The look-elsewhere effect

The search for the SM Higgs boson at LHC is performed in a mass range much wider than the expected width of the signal; the probability to observe a statistical fluctuation anywhere in the range is therefore significantly larger than the probability of observing a fluctuation at a fixed value of the mass. This effect is usually referred to as look-elsewhere effect (LEE).

In a search where the LEE is relevant, the statistical significance of the largest observed excess in the full range is quantified in terms of the global p-value, *i.e.* the probability for a background fluctuation to match or exceed the observed maximum excess anywhere in the specified mass range. In this context, the sizes of excesses at different masses are compared in terms of the local significances.

$$
\begin{aligned}
p_{\text{global}}^{\text{obs}} &= P\left(\min_{m_H} Z_{\text{local}} \geq \min_{m_H} Z_{\text{local}}^{\text{obs}} \right) \\
&= P\left(\min_{m_H} p_{\text{local}} \leq \min_{m_H} p_{\text{local}}^{\text{obs}} \right).
\end{aligned}
\tag{8.10}
$$

An intuitive characterization of the magnitude of the LEE is given by the trials factor, *i.e.* the ratio between the global p-value and the minimum local p-value, which can be interpreted as the number of independent signal hypotheses tested in the range, and approximately corresponds to the size of the range in units of the observable signal width. It is important however to note that the trials factor is not a constant, *i.e.* it depends not only on the model but also on the value of the observed minimum p-value.

When the selection requirements used in the analysis do not depend on the Higgs boson mass hypothesis probed, the global p-value can be evaluated directly from the definition (8.10) by using a large number of pseudo-datasets generated under the background only hypothesis to determine the expected distribution of the minimum local p-value in the mass range. In CMS analyses, this procedure has been used for the searches in the H $\rightarrow \gamma\gamma$ and H $\rightarrow$ ZZ $\rightarrow 4\ell$ channels.

When the selection requirements used in an analysis are dependent on the mass hypothesis tested, however, to properly account for the correlations between different hypotheses it would be necessary to generate the pseudo-datasets upstream to the hypothesis-specific requirements; this is not feasible in practice since the joint probability density of all the variables used in the hypothesis-specific requirements is not known in a simple format, and using the standard Monte Carlo event generators and detector simulation to produce these pseudo-datasets would require too large computing resources.

In the combination of all CMS searches, some of which include mass-dependent selections, the global p-value in the full search range of 110–600 GeV/c^2 is estimated by means of asymptotic formulae relating the global p-value to the mean expected number of disjoint mass regions in which the local p-values is below a chosen threshold p_{test} [104],

$$p_{global} \sim p_{local} + N(p_{local} < p_{test}) e^{-(Z^2_{local} - Z^2_{test})/2}, \qquad (8.11)$$

where p_{local} and Z_{local} are the p-value and significance of the largest excess in the range, and Z_{test} is the local significance associated to the threshold probability p_{test}. When the LEE is sizable, as is the case for the mass range considered, the mean expected number of regions can be estimated directly as the observed number of regions in the data. In order to reduce the statistical uncertainty on the estimate it is convenient to use a large value of p_{test}, so that the number of regions is maximised. In the limit case of p_{test} approaching 0.5, N is equal to the number of upcrossings, $i.e.$ deficit-to-excess transitions, observed in the mass range; as for this case $Z_{test} = 0$, equation (8.11) reduces to

$$p_{global} \sim p_{local} + N e^{-Z^2_{local}/2}. \qquad (8.12)$$

The robustness of the estimate can be checked by comparing the estimates obtained for different thresholds.

While the search for a Higgs boson has been performed in the mass range 110–600 GeV/c^2, it is interesting to quantify also the LEE when restricting the range to the low mass region preferred by the indirect constraints from electroweak observables or for which a standard model Higgs boson is not excluded by the previous direct searches. In this narrower region, the estimate of the global p-value through the number of upcrossings is not applicable because the number is only $O(1)$. However, as in this region the only two searches using mass-dependent selections are H → b$\bar{\text{b}}$ and H → WW, both characterized by a poor mass resolution; the global p-value can then be evaluated directly from pseudo-datasets in the approximation that the outcomes of those searches are totally correlated across the full mass range. This approximation is achieved by modelling the searches in those two channels at any mass hypothesis using the signal and background model at a fixed point in the range, but scaling the expected signal yield as function of the mass so that the sensitivity matches the one of the real analysis.

8.7. Modified frequentist upper limits

In order to set exclusion limits on a Higgs boson hypothesis, a test statistic q_μ is defined, dependent on the hypothesised signal rate μ. The definition of q_μ makes use of a likelihood ratio similar to the one for q_0 (eq. (8.5)), but with the signal+background model in the numerator:

$$q_\mu = -2\ln \frac{L(\text{data}, \tilde{\theta} \mid \mu, \hat{\theta}_\mu)}{L(\text{data}, \tilde{\theta} \mid \hat{\mu}, \hat{\theta})} \quad \text{with a constraint } 0 \le \hat{\mu} < \mu, \quad (8.13)$$

where the subscript μ in $\hat{\theta}_\mu$ indicates that, in this case, the maximisation of the likelihood in the numerator is done under the hypothesis of a signal of strength μ. The constrain $\hat{\mu} < \mu$ applied in the maximization at the denominator causes q_μ to be zero if the data are best described with a signal rate larger than the tested one, so that excesses are always considered to be more signal-like (*i.e.* small q_μ) than the observation of no signal ($\hat{\mu} = 0$), even when they are far too large with respect to the signal hypothesis being tested.

This definition of the test statistic differs slightly from the one used in searches at LEP and the Tevatron, where the background-only hypothesis was used in the denominator. The motivation for this choice is that in the asymptotic limit of a large number of background events, the expected distributions of q_μ under the signal+background and background-only hypotheses are known analytically [103].

Upper limits to the Higgs production cross section can be constructed using the probability to obtain a value of q_μ at least as large as the observed one under the signal+background hypothesis corresponding to that μ,

$$\text{CL}_{s+b}(\mu) = P\left(q_\mu \ge q_\mu^{obs} \mid \mu\right). \quad (8.14)$$

If one were to follow directly the Neyman construction [105], the 95% CL exclusion would be defined as $\text{CL}_{s+b}(\mu) \le 5\%$. However in the presence of a downwards fluctuation of the background this procedure can yield very tight upper limits, and by definition in 5% of the cases under the background-only hypothesis a signal of any strength $\mu \ge 0$ is excluded, irrespectively of the sensitivity of the search.

Two widely used alternatives exist to construct frequentist upper limits that are well-behaved even in the presence of strong background fluctuations: the unified approach [106], that would be obtained in this specific case relaxing the constraint $\hat{\mu} < \mu$ in the definition of q_μ, thereby resulting in upper limits or two sided intervals depending on the observed value of $\hat{\mu}$; and the modified frequentist approach [107, 108, 109], in which the condition $\text{CL}_{s+b}(\mu) \le 5\%$ is replaced by one that becomes increasingly

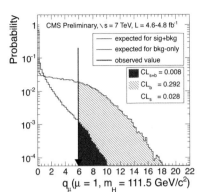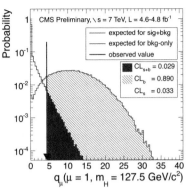

Figure 8.2. Observed value of the test statistics q_μ and expected distribution under the signal+background and background-only hypothesis, for the combination of the CMS Higgs boson searches on 2011 data at the masses of 111.5 and 127.5 GeV/c^2, for the SM hypothesis ($\mu = 1$). The probabilities CL_{s+b} and CL_b are obtained from the integral of the expected distribution above the observed value (filled and hatched areas).

more conservative for larger negative background fluctuations. In the interpretation of the results of the searches for the standard model Higgs boson at LHC it was decided to follow the second prescription, that always yields an upper limit, to provide continuity with the procedures established at LEP and Tevatron.

In the modified frequentist approach, the probability to obtain a value of q_μ at least as large as the observed one is computed also under the background-only hypothesis,

$$\mathrm{CL}_b(\mu) \;=\; P\left(q_\mu \geq q_\mu^{obs} \mid \mu = 0\right), \tag{8.15}$$

and the quantity CL_s is defined as the ratio of the two probabilities $\mathrm{CL}_s = \mathrm{CL}_{s+b}/\mathrm{CL}_b$. In this prescription the 95% CL exclusion limits are defined as $\mathrm{CL}_s(\mu) \leq 5\%$, a condition that is is always more conservative than $\mathrm{CL}_{s+b}(\mu) \leq 5\%$, and becomes increasingly so for strong downwards fluctuations of the background, *i.e.* small values of CL_b. Examples distributions of q_μ for two Higgs boson mass hypotheses are shown in Figure 8.2. The two hypotheses have similar values of CL_s but very different values of CL_b and CL_{s+b}: for the 111.5 GeV/c^2 point, where a deficit is observed, CL_b is small, and thus an even smaller value of CL_{s+b} is needed to achieve the exclusion; for the 127.5 GeV/c^2 point, instead, the CL_b is close to unity because of an excess of events, and so the limit obtained from the CL_s procedure is not far from what one would obtain using CL_{s+b}.

Recalling that $\mu = \sigma/\sigma_{SM}$, and generalizing to any confidence level, the exclusion of a standard model Higgs boson is defined by $CL_s(\mu = 1) \leq \alpha$ at the $1 - \alpha$ confidence level.

8.8. Bayesian upper limits

While the modified frequentist approach is the statistical procedure of choice for computing upper limits in the context of the Higgs boson searches, results using Bayesian inference are also computed as a check of the robustness of the interpretation of the data.

In the Bayesian approach, a posterior probability density function $f(\mu)$ for the model parameters of interest μ is obtained from the likelihood of equation (8.1) multiplied by the prior $\pi(\mu, \theta)$, after integrating out the nuisance parameters θ:

$$f(\mu) = \int \frac{1}{C} L(\text{data}, \tilde{\theta} \mid \mu, \theta) \pi(\mu, \theta) \, d\theta , \qquad (8.16)$$

where C is a normalization constant needed to enforce that $\int f(\mu) d\mu = 1$.

The choice of the prior depends on the subjective degree of belief about the values of μ and θ before the measurement. Under the assumption that the nuisance parameters are independent from the signal strength, the prior can be taken to be factorized, $\pi(\mu, \theta) = \pi(\mu) \cdot \pi(\theta)$.

In the analysis of the CMS results, as customary in high energy physics, a flat prior has been used for the signal strength μ, *i.e.* a constant $\pi(\mu)$ in the whole physically-allowed parameter space $\mu \geq 0$ up to some arbitrary boundary μ_{max}. Since the likelihood function decreases rapidly for $\mu \gg \hat{\mu}$, the posterior and the upper limit are insensitive to the choice of μ_{max} as long as it is much larger than $\hat{\mu}$.

Flat priors have been used also for the values of the nuisance parameters before the auxiliary measurements represented by $\tilde{\theta}$. Posterior probabilities for the nuisance parameters after those measurement are therefore given by

$$\tilde{p}_i(\theta_i \mid \tilde{\theta}_i) = \frac{1}{C_i} p_i(\tilde{\theta}_i \mid \theta_i) \pi(\theta_i) = \frac{1}{C_i'} p_i(\tilde{\theta}_i \mid \theta_i) , \qquad (8.17)$$

where C_i, C_i' are suitable normalizations factors. The choices of Gaussian and Poisson probabilities for p_i lead to Gaussian and gamma posteriors probabilities $\tilde{p}_i$, respectively.

Once the priors for equation (8.16) are defined, the upper limits $\mu_{95\%}$ at a 95% credible level are defined implicitly using the cumulative distribution of $f(\mu)$,

$$0.95 = \int_0^{\mu_{95\%}} f(\mu) \, d\mu . \qquad (8.18)$$

Example posterior probability densities $f(\mu)$ from the combined CMS searches are shown in Figure 8.3 for the same two Higgs boson mass hypotheses used previously in Figure 8.2.

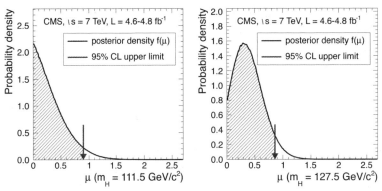

Figure 8.3. Posterior probability density functions $f(\mu)$ for the combination of the CMS Higgs boson searches on 2011 data at the masses of 111.5 and 127.5 GeV/c^2, and the corresponding upper limits on μ at 95% credible level. For the search at $m_H = 111.5$ GeV, a deficit is observed with respect to the expectations from the background-only hypothesis, so the maximum of $f(\mu)$ at $\mu = 0$; at $m_H = 127.5$ GeV/c^2, instead, the observed data lies slightly above the expectations, and so the posterior $f(\mu)$ has a maximum for a positive value of μ.

8.9. Parameter estimation

Having established the observation of a new particle in the searches for the Higgs boson, the search program has been extended naturally to the measurement of the properties of this particle.

Two broad classes of measurements are considered: the discrimination between two specific models, such as two different hypotheses on the spin or parity of this particle, and the measurement of one or more continuous parameters of a model, e.g. the mass, production cross sections or couplings. The former class, hypothesis testing, is done in a statistical framework that is identical to what done for assessing the significance of an excess or setting exclusion limits: a test statistics is defined by means of a likelihood ratio between the two different signal+background models, and the observed distribution is compared to the expectations from either of the models to compute p-values. The latter class, parameter estimation, requires instead slightly different methodology, and similarly to the case of exclusion limits different statistical techniques can be applied.

Likelihood-based estimation. This methodology relies only on the likelihood function $L(\text{data}, \tilde{\theta} \mid x, \theta)$ of the observed data in the signal region as function of the parameters of interest of the model, collectively denoted as x (*e.g.* mass, couplings). As before, systematical uncertainties enter the likelihood in the form of nuisance parameters $\tilde{\theta}$, usually constrained by external measurements $\tilde{\theta}$.

The best fit values of the model parameters, denoted as $\hat{x}$, are obtained by maximising the likelihood in x and θ. The corresponding values of the nuisance parameters are denoted as $\hat{\theta}$.

A profiled likelihood function $\lambda(x)$ is obtained from the ratio

$$\lambda(x) = \frac{L(\text{data}, \tilde{\theta} \mid x, \hat{\theta}_x)}{L(\text{data}, \tilde{\theta} \mid \hat{x}, \hat{\theta})} \quad , \tag{8.19}$$

where $\hat{\theta}_x$ is the value of the nuisance parameters maximising the numerator for the fixed value of the parameters of interest x.

Confidence intervals or regions for the values of the parameters can be determined from the asymptotic properties of the profiled likelihood function [110]: for a model with n parameters of interest, $-2 \ln \lambda$ is distributed as a chisquare with n degrees of freedom. One dimensional confidence intervals at 68% and 95% CL are therefore obtained by the conditions $-2 \ln \lambda \leq 1$ and $-2 \ln \lambda \leq 3.84$ respectively, while the equivalent conditions for the two dimensional case are $-2 \ln \lambda \leq 2.30$ and $-2 \ln \lambda \leq 5.99$.

Frequentist construction. In the frequentist paradigm, confidence regions for the parameter estimates can be obtained with the Feldman-Cousins procedure [106], using $q(x) = -2 \ln \lambda(x)$ as test statistics. With this definition, $q(x)$ is small if the predictions of the model at the parameter point x are well compatible with the data, and large otherwise. Then, a point x is defined to be inside the region for a confidence level $1 - \alpha$ if the probability to obtain a value $q(x)$ larger the observed value $q_{\text{obs}}(x)$ under the hypothesis x is larger than α,

$$R_{1-\alpha} = \left\{ x \; : \; P(q(x) > q_{\text{obs}}(x) \mid x) > \alpha \right\}. \tag{8.20}$$

The direct use of equation (8.20) to obtain confidence region intervals for the full combination of CMS Higgs boson searches is not feasible practically: it would require a very large amount of computing resources to use pseudo-experiments to extract the expected $q(x)$ distribution at all points in the parameter space. What has been done instead is to verify at a selection of some points of parameter space that the distribution of $q(x)$ is in good agreement with the asymptotic distribution described in the previous paragraph, allowing the confidence regions to be determined directly from the observed values of $q(x) = -2 \ln \lambda(x)$. Two example comparisons between the asymptotic distribution and the one extracted from pseudo-experiments is shown in Figure 8.4.

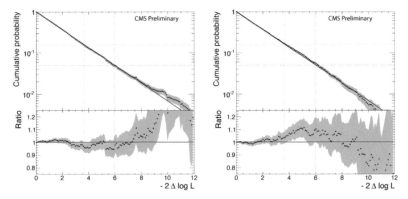

Figure 8.4. Cumulative distribution of the test statistics $q = -2\ln\lambda$ for a two-dimensional fit of the Higgs boson couplings to fermions and vector boson, for two points of the parameter space (SM on the left, and a point with reduced fermion couplings on the right). The distribution extracted from pseudo-experiments is shown by the black dots, with the statistical uncertainty represented as a green band, while the prediction from asymptotic is shown as a red solid line. In the lower panel, the ratio between the two is shown. Dashed lines show the values corresponding to 68% and 95% confidence levels.

Bayesian construction. Bayesian inference can be used to determine preferred or excluded regions of the parameter space of a model in a complementary way with respect to the frequentist construction. Similarly to what done for upper limits (sec. 8.8), a posterior probability density function in the parameter space can be determined from the likelihood function of the observed data by integrating away the nuisance parameters

$$f(x) = \int \frac{1}{C} L(\text{data}, \tilde{\theta} \mid x, \theta) \, \pi(\theta) \, d\theta \,, \qquad (8.21)$$

where $\pi(\theta)$ denotes the prior, which we assume to be independent from x.

In this approach, the point in the parameter space yielding the maximum of the posterior probability is taken to be the best estimate of the parameters of the model. Preferred parameter space regions at a given credibility level $1 - \alpha$ are defined as those containing $1 - \alpha$ of the posterior probability, after an ordering rule is chosen to decide how to select these points. In CMS, the value of the posterior probability density $f(x)$ has been used to define the ordering, *i.e.* the regions are defined as $f(x) \geq k$ where k is some value determined so that the integral of $f(x)$ in the region yields the required posterior probability. By construc-

tion, the one-dimensional intervals or two-dimensional regions constructed this way are the ones with minimal length or area among the ones
yielding the same credibility.

As for the case of upper limits, the frequentist and Bayesian approach
are in principle independent, but in practice yield similar results, confirm-
ing that the physics interpretation of the results is robust with respect to
the choice of the statistical methodology. A side-by-side comparison of
the results of the likelihood-based and Bayesian approaches for the combined measurement of the mass and signal cross section from the H $\rightarrow$ $\gamma\gamma$
are shown in Figure 8.5.

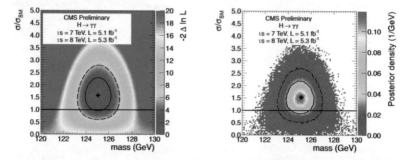

Figure 8.5. Comparison of the 68% and 95% CL regions for a bidimensional
analysis of the signal mass and production cross section for the H $\rightarrow$ $\gamma\gamma$ decay
mode. The result on the left is from a likelihood-based analysis approximating
the frequentist construction, the one on the right from a Bayesian analysis.

Chapter 9
Results of the CMS searches for a standard model Higgs boson

In this chapter, the results of the searches for a standard model Higgs boson at CMS are presented. After an overview of the sensitivities, the results from the searches in the individual decay modes are introduced first, followed by the combined analysis of all channels together to achieve the maximal sensitivity.

The list of all channels covered in the CMS searches for a SM Higgs boson is summarized in Table 9.1, including the Higgs boson mass range considered and the approximate resolution on the mass of a reconstructed signal. In the table, individual references to the preliminary results are also given.

A preliminary combination of all the results with the summer 2012 dataset is described in reference [111]. The results in the low mass range, for the five most sensitive analyses that include both 2011 and 2012 data, are also published unmodified in reference [23]. The results described in this chapter are taken mostly from the preliminary combination, as it includes the full mass range and all the channels.

The sensitivity of each channel to a SM-like Higgs boson can be well quantified in terms of the expected upper limit on the signal strength parameter $\mu = \sigma/\sigma_{SM}$. The expected upper limit, $\mu_{95\%}$, is also an indication of how much each individual channel contributes to the combined result, as the weight of each channel is approximately proportional to $1/\mu_{95\%}^2$.

The expected upper limits for each of the individual channels are shown in Figure 9.1. At low mass, the channels with the largest sensitivities are H $\rightarrow$ $\gamma\gamma$ and H $\rightarrow$ ZZ, with H $\rightarrow$ WW very close to it. Nonetheless, also the other modes contribute significantly to the overall result, as clearly illustrated by the difference between the expected limit for the full combination and that of the most sensitive channel alone. In addition to improving the overall sensitivity, the searches in multiple channels are extremely important to provide robustness against potential experimental issues in individual final states, and to evaluate the consistency of any

Channel	m_H range (GeV/c^2)	Luminosity (fb^{-1}) $\sqrt{s} = 7\,\text{TeV}$	$\sqrt{s} = 8\,\text{TeV}$	m_H resolution	Reference
$VH, H \to b\bar{b}$	110–135	5.0	5.1	10%	[112]
$t\bar{t}H, H \to b\bar{b}$	110–140	5.0	–	–	[113]
$H \to \tau\tau$	110–145	4.9	5.1	20%	[114]
$WH, H \to \tau\tau$	110–140	4.9	–	20%	[115]
$ZH, H \to \tau\tau$	110–160	5.0	–	20%	[116]
$H \to \gamma\gamma$	110–150	5.1	5.3	1–2%	[117]
$H \to WW \to 2\ell 2\nu$	110–600	4.9	5.1	20%	[118, 119]
$H \to WW \to \ell\,\nu q\bar{q}$	180–600	5.0	5.1	20%	[120, 121]
$WH \to 3\ell 3\nu$	110–200	4.9	–	–	[122]
$VH \to jj\,2\ell 2\nu$	118–180	4.9	–	–	[123]
$H \to ZZ \to 4\ell$	110–600	5.0	5.3	1–2%	[124]
$H \to ZZ \to 2\ell 2\tau$	190–600	5.0	5.3	10–15%	[124]
$H \to ZZ \to 2\ell 2q$	130–164 / 200–600	4.9	–	3% / 3%	[125]
$H \to ZZ \to 2\ell 2\nu$	200–600	4.9	5.1	7%	[126]

Table 9.1. Summary of the SM Higgs boson searches performed at CMS, grouped by dominant decay mode; the WH search with τ_h in the final state also includes contributions from WH $\to$ 3W with W $\to \tau_h\,\nu$.

observed excess with the hypothesis of a SM Higgs boson signal (or of any alternative model). The sensitivity of the channels can also be quan-

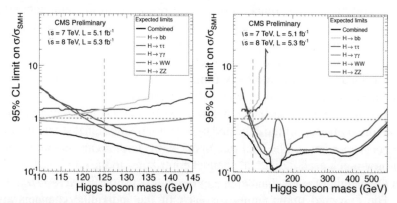

Figure 9.1. Expected upper limits at 95% confidence level on the signal strength parameter $\mu = \sigma/\sigma_{SM}$, for the five decay modes and their combination. The sharp changes in sensitivity as function of m_H in some channels are due to changes in the event selection or in the training of multivariate classifiers, or for H $\to$ b$\bar{\text{b}}$ and H $\to$ $\tau\tau$, by changes in the list of modes analyzed at each mass hypothesis.

tified in terms of the significance of the excess expected for a SM Higgs boson signal (Figure 9.2). The hierarchy of the channels is similar but not identical to the one obtained from the expected upper limits, as the subchannels or bins with low statistics but high purity have a different weight in the two computations. The median expected significance for a SM Higgs boson signal from the combination of all the channels is above 5σ for Higgs boson mass hypotheses above $122\,\mathrm{GeV}/c^2$.

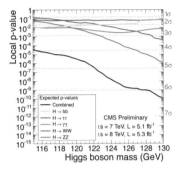

Figure 9.2. The median expected p-value for observing an excess at mass m_H assuming that the SM Higgs boson with that mass exists, in the range 110–$130\,\mathrm{GeV}/c^2$. Expectations are shown for the full combination and for each of the five decay modes explored at CMS.

Presentation of the results. Unless stated otherwise, the following conventions are used in presenting these results: the observed values are shown by a solid line, with markers corresponding to the actual m_H hypotheses probed; a dashed line indicates the median of the expected results for the background-only hypothesis; the green and yellow bands indicate the ranges in which the measured values are expected to reside in at least 68% and 95% of all experiments under the background-only hypothesis.

When presenting local p-values p_0, the expectations from the background-only hypothesis are not shown, as by definition p_0 is uniformly distributed between 0 and 1; instead, for each value of m_H, the median expected p-value under the hypothesis of standard model Higgs boson of mass m_H is shown.

For practical convenience, in most cases limits and p-values are computed from the asymptotic properties of the profile likelihood, which was shown in the past to agree to $O(10\%)$ or better with the full frequentist procedure using pseudo-datasets.

9.1. Results from individual searches

9.1.1. H → $\gamma\gamma$ searches

The upper limits obtained in the search for a standard model Higgs boson in the H → $\gamma\gamma$ channel are shown in Figure 9.3. The H → $\gamma\gamma$ analysis

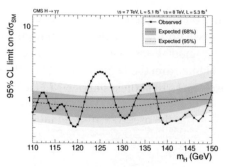

Figure 9.3. 95% CL upper limits on the signal strength parameter $\mu = \sigma/\sigma_{SM}$ for the H $\rightarrow \gamma\gamma$ search, as function of the Higgs boson mass.

is characterized by a good mass resolution and by a large and featureless background. As a natural consequence of this, positive and negative fluctuations are expected in the results; also, since the normalization of the background is determined only a posteriori from the observed diphoton spectrum, by design positive and negative excursions approximately cancel out in the full range.

In order to better assess the size of the observed excesses and deficits, it is useful to represent the results in terms of the value of the $\mu = \sigma/\sigma_{SM}$ that best describes the data; for this search, the largest excesses at about 125 GeV/c^2 corresponds to a σ/σ_{SM} value of 1.6 ± 0.4 (Figure 9.4, left). In this presentation, the constraint $\mu \geq 0$ is relaxed, to allow representing also the deficits of events with respect to the background-only expectations.

The statistical significance of the positive excursions is quantified in terms of p-values, as described in Section 8.5. The excess at diphoton masses around 125 GeV/c^2, that appears consistently in the 7 and 8 TeV datasets, has a local significance of 4.1σ (Figure 9.4, right), larger than the median expected for a SM Higgs boson signal (2.8σ). The oscillations in the expected p-value are a consequence of the definition used: the parameters of the signal model are taken from a fit to the data with μ fixed to unity, resulting in positive or negative oscillation when the data would be better described by a value of μ significantly larger or smaller than one, and thus mirroring to a smaller scale the oscillations in the observed p-value.

Since the search is performed in a mass range much larger than the observable width of a signal, a sizable look-elsewhere effect is present; when this is taken into account, the probability of observing an excess with a local significance of 4.1σ anywhere in the spectrum under the background-only hypothesis is found to be about $7.2 \cdot 10^{-4}$ (3.2σ).

The statistical significances of the other excesses and deficits in the plot of Figure 9.4 (left) has also been evaluated[1], and are all much smaller, in the 1–2σ range. The largest deficit, at about 147.5 GeV/c^2, has a local significance of 2.7σ, and after accounting for the look-elsewhere effect a deficit of that size or larger somewhere in the spectrum is expected in about 8% of the background-only datasets. The probability of such a deficit is likely even a bit larger in the presence of a signal somewhere in the spectrum, as the determination of the background from sidebands at Higgs boson mass hypotheses away from the signal will be biased upwards by the signal events, although a quantitative evaluation of this effect has not been done.

The result from the nominal analysis is confirmed by cross-check analyses based on different approaches, one not relying on multivariate methods for photon selection and diphoton categorization, and one combining also the mass variable in a multivariate classifier and obtaining the background from sidebands in the diphoton mass.

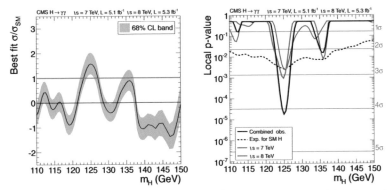

Figure 9.4. Observed best fit signal strength $\hat{\mu} = \sigma/\sigma_{SM}$ (left) and local p-value p_0 (right) for the H $\rightarrow \gamma\gamma$ search, as function of the Higgs boson mass. The dashed blue line in the right plot represents the expected local p-value $p_0(m_H)$ should a SM Higgs boson with mass m_H exist.

9.1.2. H $\rightarrow$ b$\bar{\text{b}}$ and H $\rightarrow \tau\tau$ searches

The searches in the two decay modes H $\rightarrow \tau\tau$ and H $\rightarrow$ b$\bar{\text{b}}$ have similar sensitivity, and are both characterized by a mass resolution comparable to the size of the probed range. The observed limits obtained from the analysis of the results from these searches are close to the expected ones, with H $\rightarrow$ b$\bar{\text{b}}$ exhibiting an excess of about one standard deviation and

---

[1] For deficits, the definition of q_0 from eq. (8.5) has to be modified by dropping the requirement $\mu \geq 0$.

H → ττ a deficit of similar size, as shown in Figure 9.5. The results shown in Figure 9.5 do not include the channels for which only the 2011 data was analyzed (t$\bar{\text{t}}$H, H → b$\bar{\text{b}}$ and VH, H → ττ), which are included in the full combination but anyway contribute less to the sensitivity. The discontinuities in the observed limit for the H → b$\bar{\text{b}}$ case are due to changes in the selection criteria corresponding to different trainings of the multivariate classifier.

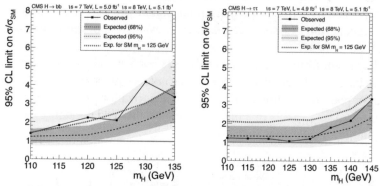

Figure 9.5. 95% CL upper limits on the signal strength parameter $\mu = \sigma/\sigma_{SM}$ for the searches in the H → b$\bar{\text{b}}$ (left) and H → ττ (right) decay modes. The results do not include the channels for which only the 2011 data was analyzed (t$\bar{\text{t}}$H, H → b$\bar{\text{b}}$ and VH, H → ττ), which anyway contribute little to the sensitivity. The median expected limit at each m_H hypothesis, should a SM Higgs boson with mass 125 GeV/c^2 exist, is also shown as a blue dotted line.

9.1.3. H → WW searches

The expected and observed upper limits from the H → WW searches are shown in Figure. 9.6. The H → WW channel has the largest sensitivity in the 150–190 GeV/c^2 mass range, and totally dominates the search in the interval between $2m_W$ and $2m_Z$ where decays to on-shell bosons are possible for WW but not for ZZ. The final states contributing more to the sensitivity of the H → WW search are the dilepton states with zero or one jet, due to the larger cross section for the gluon fusion production mode compared to the vector boson fusion and associated production modes.

As for H → b$\bar{\text{b}}$ and H → ττ, the absence of a narrow signal peak in this mode causes the outcome of the searches at different m_H values to be strongly correlated, and so any signal is expected to manifest as a broad excess, and a similar behaviour is expected also for statistical fluctuations in the background. However, unlike those two channels, the branching fraction of the decay into WW pairs is very strongly dependent on m_H below the $2m_W$ threshold, and so are therefore the expected and observed limits in terms of σ/σ_{SM}.

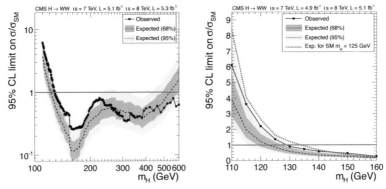

Figure 9.6. 95% CL upper limits on the signal strength parameter $\mu = \sigma/\sigma_{SM}$ for the H $\rightarrow$ WW searches, as function of the Higgs boson mass. The result on the left, for the full mass range, is for the combination of all H $\rightarrow$ WW production and decay topologies. The figure on the left, instead, showing the low mass region, is only for the H $\rightarrow$ WW $\rightarrow$ $2\ell 2\nu$ decay mode in the final states with zero, one or two jets; the median expected limit at each m_H hypothesis, should a SM Higgs boson with mass 125 GeV/c^2 exist, is also shown as a blue dotted line.

In this search, an excess of signal-like events compared to the background predictions is observed in the whole m_H region below about 250 GeV/c^2, with modulations coming from the changes in the selection criteria and in the trainings of the multivariate classifier. The statistical significance of the excess is about two standard deviations irrespectively of m_H, since by definition it does not depend on the expected signal rate (Figure 9.7, left). Conversely, because of the strong dependence of $\sigma \times BR_{WW}$ from m_H, the interpretation of the excess in terms of a SM Higgs boson signal is clearly disfavoured for Higgs boson mass hypotheses above about 135 GeV/c^2, as indicated both from the exclusion limits (Figure 9.6, right) and the best fit σ/σ_{SM} values (Figure 9.7, right).

9.1.4. H $\rightarrow$ ZZ searches

The searches in the H $\rightarrow$ ZZ decay mode at low mass are driven almost exclusively by the H $\rightarrow$ ZZ $\rightarrow$ 4ℓ, as the sensitivity of the H $\rightarrow$ ZZ $\rightarrow$ $2\ell 2q$ search at low mass is limited by the larger Z + jets background. At high mass, instead, the searches in the H $\rightarrow$ ZZ $\rightarrow$ $2\ell 2\nu$ and H $\rightarrow$ ZZ $\rightarrow$ $2\ell 2q$ play a larger role; as the natural width of the Higgs boson increases proportionally to m_H^4, faster than the experimental resolution on reconstructed quantities ($\Delta p_T/p_T \propto p_T$, so $\Delta m \propto m^2$), for heavy Higgs

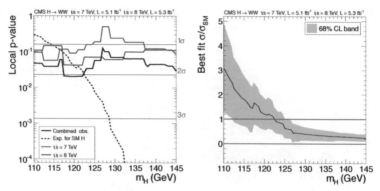

Figure 9.7. Observed local p-value p_0 (left) and best fit signal strength $\hat{\mu} = \sigma/\sigma_{SM}$ (right) for the H $\to$ WW $\to$ 2ℓ2ν searches, as function of the Higgs boson mass. The dashed blue line in the left plot represents the expected local p-value $p_0(m_H)$ should a SM Higgs boson with mass m_H exist.

boson hypotheses the observable width of a signal becomes the same in all channels.

Upper limits on the Higgs boson production from the combination of the searches in the H $\to$ ZZ decay mode are shown in Figure 9.8; in general the sensitivity as function of m_H reflects the dependency of $\sigma_{SM} \times BR_{H \to ZZ}$ on the Higgs boson mass, except that at high mass the decrease in cross section is partially compensated by the increase in signal efficiency and discrimination power.

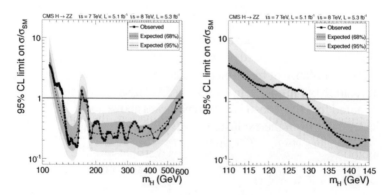

Figure 9.8. 95% CL upper limits on the signal strength parameter $\mu = \sigma/\sigma_{SM}$ for the H $\to$ ZZ searches, as function of the Higgs boson mass.

Statistical significance of the excess of events observed in data at low mass and at high values of the kinematic discriminator is 3.2σ for the nominal two-dimensional analysis using m_{ZZ} and K_D (Figure 9.9). The minimum of the p-value is at $m_H \sim 125.5\,\mathrm{GeV}/c^2$, and the corresponding best fit value of σ/σ_{SM} is $0.7^{+0.4}_{-0.3}$. The expected significance for a SM Higgs boson of mass $125.5\,\text{GeV}/c^2$ is 3.8σ. For the cross-check analysis relying only on m_{ZZ}, less sensitive than the nominal one, the statistical significance of the observed excess is about 2.1σ.

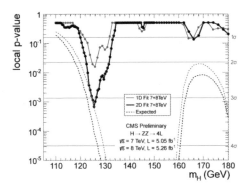

Figure 9.9. Observed local p-value p_0 for the $H \rightarrow ZZ \rightarrow 4\ell$ search as function of the Higgs boson mass for the nominal analysis (black) and for the one-dimensional analysis using only m_{ZZ} and no angular variables (blue). The dashed black and blue lines in the plot represents the expected local p-value $p_0(m_H)$ for the two analyses, should a SM Higgs boson with mass m_H exist.

9.2. Results from the combination of all CMS searches

Information from all individual searches is combined and analyzed using statistical inference as described in Chapter 8. Limits will be described first, following by a statistical characterization of the significance of the observed excess and its compatibility with the prediction of a SM Higgs boson.

9.2.1. Exclusion limits

For setting upper limits for the full combination, the nominal result is computed with the modified frequentist CL_s approach relying on the asymptotic properties of the test statistics distribution. The upper limit on the signal strength modifier σ/σ_{SM} for the combination of the 2011 and 2012 searches is shown in Figure 9.10: CMS is sensitive to a Higgs boson of SM cross section in the full mass range probed, 110–$600\,\text{GeV}/c^2$, and would be sensitive to a Higgs-like boson with reduced yield down to about 0.2 times the standard model yield in a wide mass range.

A magnification of the CL_s limits in the low mass region is shown in Figure 9.11 (left), together with the values of CL_s for the standard model Higgs signal hypothesis (right). A standard model Higgs mass hypotheses is excluded at the 95%, 99%, and 99.9% if the corresponding

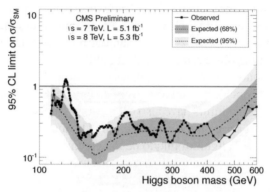

Figure 9.10. 95% CL upper limits on the signal strength parameter $\mu = \sigma/\sigma_{SM}$ from the combination of the 2011 and 2012 CMS searches[111], as function of the Higgs boson mass.

observed CL_s value is below the thresholds of 0.05, 0.01, and 0.001, shown as red lines in Figure 9.11 (right); expected exclusions can be obtained similarly from the median expected CL_s value.

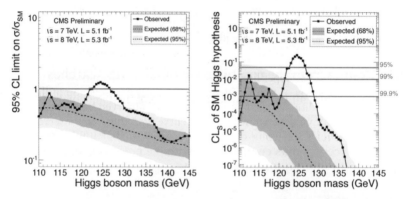

Figure 9.11. 95% CL upper limits on the signal strength parameter $\mu = \sigma/\sigma_{SM}$ (left) and CL_s values for the SM hypothesis $\mu = 1$ (right), as function of the Higgs boson mass, for the combination of all CMS searches on 2011 and 2012 data [111].

The outcome of the combination of all CMS Higgs searches is that the standard model Higgs boson is excluded at 95% CL in the mass ranges 110–122.5 and 127–600 GeV/c^2, and at 99% CL in the ranges 110–112, 113–121.5 and 128–600 GeV/c^2. In general, the differences between the observed and expected limits are consistent within the statistical fluctuations, as the observed limits are within the 68% and 95% bands of the expected outcomes. However, at small m_H the observed limit is weaker than the expectations under the background-only hypothesis due to an overall excess of events.

Other statistical methods.   While the upper limits for the combination of the 2011 and 2012 dataset has been made only using the asymptotic CL_s approximation, for the 2011 dataset alone the full CL_s procedure has also been used, relying on ensembles of pseudo-datasets to determine the expected distribution of the test statistics. To further assess the invariance of the physical conclusions under different statistical approaches, for that dataset the observed limits have also been computed using a Bayesian approach with a flat prior on the cross section. The three exclusion limits on σ/σ_{SM} are shown in figure found to be in agreement within 10% or better (Figure 9.12), as expected from past experiences at LHC and Tevatron.

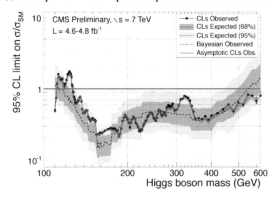

Figure 9.12. 95% CL upper limits on the signal strength parameter $\mu = \sigma/\sigma_{SM}$ from the combination of the 2011 CMS searches[21], as function of the Higgs boson mass. In addition to the nominal CL_s limits, observed limits from asymptotic formulae for CL_s are shown as a red curve, and Bayesian observed limits as blue open circles and a blue dashed line.

9.2.2. Statistical significance of the observed excess

In order to quantify the statistical significance of the low mass excess, local p-values are computed from the asymptotic properties of the test statistics q_0 (eq. 8.8). The minimum local p-value is observed for $m_H = 125.5\,\text{GeV}/c^2$, where the local significance is evaluated to be 4.9σ, in fair agreement with the median expected significance for a Higgs boson signal of that mass, 5.9σ. The excess appears consistently for the 7 and 8 TeV, with local significances of 3.0σ and 3.8σ respectively (Figure 9.13).

The two channels giving the largest contribution to the excess are the two modes with the largest sensitivity and the best mass resolution, $H \rightarrow \gamma\gamma$ and $H \rightarrow ZZ \rightarrow 4\ell$ modes. The observed local significances at $m_H = 125.5\,\text{GeV}/c^2$ are 4.0σ for the $H \rightarrow \gamma\gamma$ search and 3.2σ for the $H \rightarrow ZZ \rightarrow 4\ell$ search; the corresponding expected significances for the two modes under the hypothesis of a SM Higgs boson of mass $125.5\,\text{GeV}/c^2$

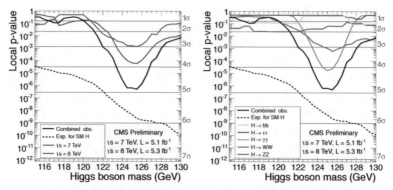

Figure 9.13. The observed local p-value p_0 as a function of the Higgs boson mass from the combination of the 2011 and 2012 datasets[111]. The separate results from the two data taking periods and from the individual Higgs boson decay modes are also shown, in the left and right plots respectively.

are $2.8\,\sigma$ and $3.8\,\sigma$, respectively. The local significance of the excess observed in the combination of these two modes alone is $5.0\,\sigma$, for an expected significance of $4.7\,\sigma$.

The $H \rightarrow WW \rightarrow 2\ell2\nu$ search channel has a good signal sensitivity but a poor mass resolution. As described previously, a $1.5\,\sigma$ excess is observed in this mode, slightly smaller than the median expected for a SM Higgs boson signal, $2.5\,\sigma$. When added to the combination of the $H \rightarrow \gamma\gamma$ and $H \rightarrow ZZ \rightarrow 4\ell$ searches, the observed local significance becomes $5.1\,\sigma$, and the expected $5.2\,\sigma$.

The searches in the fermionic decay modes $H \rightarrow b\bar{b}$ and $H \rightarrow \tau\tau$ have both a poor mass resolution and a smaller sensitivity compared to the bosonic modes: the expected local significances for a SM Higgs of mass $125.5\,\mathrm{GeV}/c^2$ are $1.9\,\sigma$ for $H \rightarrow b\bar{b}$ and $1.5\,\sigma$ for $H \rightarrow \tau\tau$. The results of the $H \rightarrow b\bar{b}$ search exhibit only a tiny excess of $0.2\,\sigma$ ($0.7\,\sigma$ from the VH modes alone, excluding $t\bar{t}H$), and a deficit of events is observed in the $H \rightarrow \tau\tau$ mode. Therefore, when including also these channels the median expected significance increases ($5.9\,\sigma$) but the the observed excess is reduced ($4.9\,\sigma$).

The results for each decay mode and their combinations are summarized in Table 9.2.

Look-elsewhere effect.   The global significance of the observed excess in the full mass range 110–$600\,\mathrm{GeV}/c^2$ is $4.0\,\sigma$, as estimated using the number of deficit-to-excess transitions observed in the data (Section 8.6).

Several different considerations could be invoked to restrict the range of interest only to lighter Higgs boson masses: constraints from elec-

Channel	expected	observed
H $\rightarrow$ ZZ $\rightarrow$ 4ℓ	3.8σ	3.2σ
H $\rightarrow$ $\gamma\gamma$	2.8σ	4.0σ
H $\rightarrow$ WW $\rightarrow$ 2ℓ2 ν	2.5σ	1.5σ
H $\rightarrow$ b$\bar{\text{b}}$	1.9σ	0.2σ
H $\rightarrow$ $\tau\tau$	1.5σ	deficit
H $\rightarrow$ $\gamma\gamma$, ZZ	4.7σ	5.0σ
H $\rightarrow$ $\gamma\gamma$, ZZ , WW	5.2σ	5.1σ
H $\rightarrow$ b$\bar{\text{b}}$, $\tau\tau$	2.4σ	deficit
all modes	5.9σ	4.9σ

Table 9.2. Observed local significances for the search at a Higgs boson mass hypothesis of 125.5 GeV/c^2, and median expected significances for a SM Higgs boson of that mass, for the different decay modes and their combinations.

troweak precision observables; observed limits from searches at LEP, Tevatron, ATLAS or even past CMS results; the compatibility with a supersymmetric extension to the standard model. As the result does not depend strongly on the choice of the range, for illustration the 115–130 and 110 − 145 GeV/c^2 intervals have been used as representatives of all these possible restricted ranges to compute the global significance.

In a restricted mass range the method based counting deficit-to-excess transitions can no longer be used, but the expected distribution of the minimum local p-values can be determined directly in an ensemble pseudo-datasets generated under the background-only hypothesis[2]. The expected distribution of minimum local p-values can then be used to connect a minimum local p-value to a global p-value, at least for moderate values of the significance. In order to extrapolate this procedure to the very small observed p-value 5.5 · 10^{-7}, as it is not computationally feasible to generate a matching number of pseudo-datasets, the asymptotic relationship connecting local and global p-values is used, $p_{\text{global}} \sim p_{\text{local}} + N e^{-Z_{\text{local}}^2/2}$. In this context, the otherwise unknown parameter N is fitted from the pseudo-datasets for smaller values of the significance.

After the correction for the look-elsewhere effect in the 115–130 GeV/c^2 mass range, the observed global significance is 4.5σ; the corresponding number for the 110 − 145 GeV/c^2 is similar, 4.4σ.

---

[2] For this purpose, the analyses in final states with mass-dependent selections, WW and b$\bar{\text{b}}$, are approximated using the outcome of the search at a fixed m_{H} value the center of the range. The same treatment is applied to the H $\rightarrow$ $\tau\tau$ search, as the mass resolution is anyway poor compared to the search range considered.

The large value of the observed significance and the consistent pattern with which the excess manifests itself in the different running periods and channels lead to the interpretation of the excess as the existence of a new heavy state with mass close to $125.5\,\mathrm{GeV}/c^2$. The observation of the $\gamma\gamma$ and 4ℓ decay modes indicate that this state is a neutral boson; furthermore, the spin of the particle cannot be 1, as in that case the $\gamma\gamma$ decay is forbidden [127, 128].

9.2.3. Compatibility with the SM Higgs hypothesis

The compatibility of an excess with the expectations from a standard model Higgs boson signal is quantified in terms of the best fit value $\hat{\mu}$ of the signal strength modifier $\mu = \sigma/\sigma_{SM}$. As shown in Figure 9.14, the size of the observed excess for m_{H} around $125\,\mathrm{GeV}/c^2$ is compatible with the SM expectations, while in the rest of the low mass region it is mostly compatible with zero. The overall best fit value for a Higgs boson mass hypothesis of $125\,\mathrm{GeV}/c^2$ is found to be $\mu = 0.80 \pm 0.22$.

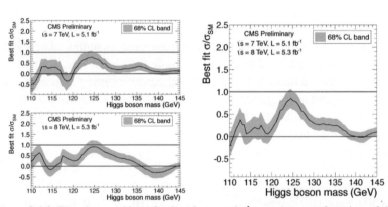

Figure 9.14. The observed best fit signal strength $\hat{\mu} = \sigma/\sigma_{SM}$ as function of the Higgs boson mass for the combination of all CMS searches. Left: 7 and 8 TeV data separately, in the top and bottom panels; Right: full combination.

In addition to the consistency of the overall result with the SM Higgs boson hypothesis, the self-consistency of the result is evaluated by comparing the observed $\hat{\mu}$ values in the individual channels. A first comparison is done by separating channels that differ in the decay mode or in the production topology, as shown in Figure 9.15. The sensitivity of the individual modes is limited, but in general a fair agreement is observed between the results.

A more quantitative result can be obtained by computing a generalized chisquare from the profiled likelihood as function of μ in each of the

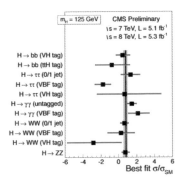
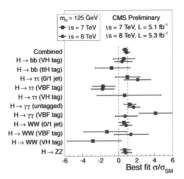

Figure 9.15. The observed best fit signal strength $\hat{\mu} = \sigma/\sigma_{SM}$ in different channels, for the combination of the 2011 and 2012 datasets (left) and for the two separately (right). The vertical green band in the left plot shows the overall $\hat{\mu}$ value 0.80 ± 0.22. The horizontal error bars on the individual results indicate the $\pm 1\sigma$ uncertainties, and include both statistical and systematical contributions.

channels,

$$\chi^2 = -2 \sum_{i=1}^{n} \ln \frac{L_i(\text{data}|\hat{\mu})}{L_i(\text{data}|\hat{\mu}_i)} \tag{9.1}$$

where $\hat{\mu}_i$ is the best fit value from the i-th channel alone. Neglecting the correlation across the channels, this variable is expected to be distributed approximately as a χ^2 with a number of degrees of freedom equal to the number of channels minus one. The application of equation (9.1) to the full 2011 and 2012 datasets yields a global χ^2 value of 13.5, well compatible with the expectations for 10 degrees of freedom: the largest contribution is from the VBF modes of H $\to$ $\tau\tau$ ($\Delta\chi^2 = 6.8$), followed by the H $\to$ $\gamma\gamma$ modes (both with $\Delta\chi^2 \sim 2$). A similar compatibility test can be done also with respect to the SM Higgs boson hypothesis, by using $\mu = 1$ instead of $\hat{\mu}$ in the numerator of eq. (9.1): the corresponding χ^2 value obtained is 14.1, compatible with the expectations for 11 degrees of freedom.

The compatibility tests can also be performed by combining channels by decay mode, inclusively on the production topology, or by combining by production topology inclusively on the decay mode (Figure 9.16). In this case, the $\hat{\mu}$ values from the sub-combinations are measured more accurately, yielding a more precise test of the compatibility with a SM Higgs boson; however the interpretation of any possible deviation in terms of alternative models is now obscured by the use of SM relationships across production or decay modes within the sub-combinations. Also for these tests the channels are found to be in agreement with the uncertainties.

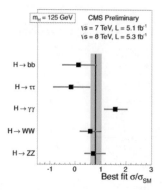 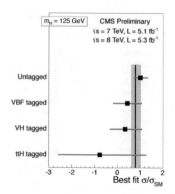

Figure 9.16. The observed best fit signal strength $\hat{\mu} = \sigma/\sigma_{SM}$ when combining channels by decay mode (left) or production topology (right). The vertical green band hows the overall $\hat{\mu}$ value 0.80 ± 0.22, while horizontal error bars on the individual results indicate the $\pm 1\sigma$ uncertainties, and include both statistical and systematical contributions.

As the signal from the newly observed heavy boson is fairly compatible with the predictions of a SM Higgs boson, it is possible to perform some first measurements of its properties from the results of the search analyses, relying on the SM Higgs predictions for combining different channels and for modelling the kinematic distributions and the detector response to this signal.

From a strict perspective, since all the theoretical predictions used in the analyses to model the signal are valid only for a SM Higgs boson, all the measurements that can be extracted from those analyses can only be considered as tests of the SM Higgs hypothesis. This is manifestly clear *e.g.* when considering the best fit σ/σ_{SM} values extracted from sub-combinations by decay mode inclusively on the production mode, since a non-standard Higgs-like boson with different decay branching fraction will almost necessarily have also different production cross sections in the different topologies. However, more subtle dependencies on the model can also be there when considering individual production topologies and decay modes: *e.g.* the light Higgs boson of the minimal supersymmetric extension to the standard model (MSSM) has, in some regions of the parameter space, an inclusive production cross section in the gluon fusion similar to that of a SM Higgs boson of the same mass, but a different p_T spectrum[129], and thus a different experimental acceptance. However, if the signal observed in the data does not depart too much from the SM Higgs boson predictions, then the measurements done relying on some standard model assumptions can be expected to be a fair estimate of the real deviations from the SM predictions, as the bias introduced by the model assumptions will be a second order correction to the result.

9.3. Measurement of the mass

The mass of the boson is measured from the invariant mass distribution of the decay products, in the two channels where this can be performed accurately: $H \rightarrow \gamma\gamma$ and $H \rightarrow ZZ \rightarrow 4\ell$. Both for the SM Higgs boson and for most alternative models proposed in the literature, the natural width of a light Higgs-like boson is negligible with respect to the experimental resolution, and therefore the mass is unambiguously defined also from the theoretical point of view.

In the presence of background and for a small sample size, some dependency on the model assumptions will enter inevitably in the measurements: the events in data contribute to the measurement with a weight approximatively proportional to the expected signal purity $S/(S + B)$, and so the use of a signal model other than the true one will result in an incorrect weighting of the events; in this specific case, *e.g.* the relative weight of the different $H \rightarrow \gamma\gamma$ categories depends on the assumed p_T spectrum of the Higgs boson and the hierarchy of the cross sections in the different production topologies, while the weight of each $H \rightarrow ZZ \rightarrow 4\ell$ event depends on the assumed angular distribution of the Higgs boson decay products.

The measurement is first performed separately in the three final states $H \rightarrow ZZ \rightarrow 4\ell$, $H \rightarrow \gamma\gamma$ without VBF tag, and $H \rightarrow \gamma\gamma$ with VBF tag, from a fit to the data without relying on the SM Higgs prediction for the expected signal yield. The results are then combined, either letting the signal yield vary independently in each of the modes (Figure 9.17, left) or constraining them to a single parameter relying on the SM predictions for the ratios of the cross sections in the different topologies and for the ratio of the branching fractions (Figure 9.17, right). The former combination, less dependent on the SM Higgs boson assumptions, is taken as nominal result. The good agreement between the two results is an indication of the robustness of the measurement.

In order to separate the purely statistical uncertainties on the mass measurement from the systematical ones, the one-dimensional likelihood scan as function of mass is also performed without profiling the nuisance parameters and the signal strength (Figure 9.18). The systematical uncertainty is then determined by the subtraction in quadrature between the total uncertainty, determined from the scan with the profiled likelihood, and the statistical one, determined from the non-profiled likelihood.

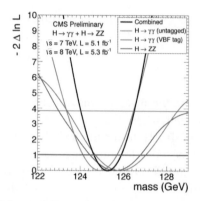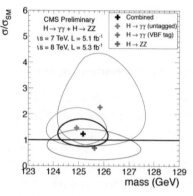

Figure 9.17. Left: likelihood scan as function of the mass for the three analyzed final states, and the model-independent combination of the three; the horizontal lines at 1.0 and 3.84 indicate the threshold levels defining for the $\pm 1\sigma$ and 95% CL intervals. Right: best fit values and 68% CL regions for the two-parameter fits of mass and signal normalization in each final state analyzed and in their model-dependent combination, where the SM Higgs predictions are used for the ratios of the signal yields in the three modes.

The result of the mass measurement from the combination of the high resolution channels is $m = 125.3 \pm 0.4^{(\text{stat})} \pm 0.5^{(\text{syst})}$ GeV/c^2. The dominant source of systematical uncertainty is from the extrapolation of the ECAL energy scale from the $Z \rightarrow e^+e^-$ calibration point to the $H \rightarrow \gamma\gamma$ decay, from possible data-to-simulation differences in the relative response of the detector to electrons and photons, and in the extrapolation of the calibrations from a di-object mass of 91 GeV/c^2 to 125 GeV/c^2.

9.4. Probing the couplings of the observed boson

9.4.1. Test of custodial symmetry

Electroweak symmetry breaking via the Higgs mechanism sets a well-defined ratio for the couplings of the Higgs boson to the W and Z bosons, $g_{\text{HWW}}/g_{\text{HZZ}}$, protected by the custodial symmetry. This ratio is also preserved in a large set of alternative theoretical models that could accommodate a Higgs-like boson with mass around 125 GeV/c^2.

The dominant production mechanism populating the inclusive pp $\rightarrow$ H $\rightarrow$ ZZ and untagged pp $\rightarrow$ H $\rightarrow$ WW search channels is gg $\rightarrow$ H. Therefore the ratio of event yields in these channels provides a natural test of the custodial symmetry.

To quantify such consistency, the search results are interpreted by parameterizing the expected signal yield as function of two quantities, μ_{ZZ} and $R_{\text{W/Z}}$. The expected H $\rightarrow$ ZZ $\rightarrow$ 4ℓ event yield is scaled with

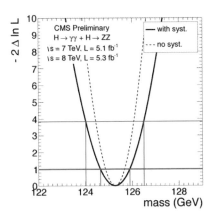

Figure 9.18. Likelihood scan as function of the mass for the model-independent combination of the H $\rightarrow$ ZZ $\rightarrow$ 4ℓ and H $\rightarrow$ $\gamma\gamma$ searches. The profiled likelihood function is shown as a solid curve, and the non-profiled one as a dashed curve. The horizontal lines at 1.0 and 3.84 indicate the threshold levels defining for the $\pm 1\sigma$ and 95% CL intervals.

respect to the SM Higgs boson predictions by μ_{ZZ}, while the expected untagged H $\rightarrow$ WW $\rightarrow$ 2ℓ2ν event yield is scaled by $R_{W/Z} \cdot \mu_{ZZ}$. The mass of the observed state is fixed to 125 GeV/c^2, but the result does not depend strongly on this choice.

A scan of the likelihood as function of $R_{W/Z}$, while profiling all other nuisances and the signal strength modifier μ_{ZZ}, yields $R_{W/Z} = 1.0^{+1.1}_{-0.6}$ (Figure 9.19).

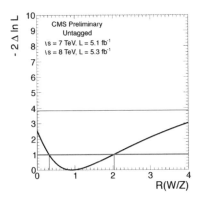

Figure 9.19. Likelihood scan as function of $R_{W/Z}$ for the combination of the H $\rightarrow$ ZZ and non-VBF H $\rightarrow$ WW searches. The horizontal lines at 1.0 and 3.84 indicate the threshold levels defining for the $\pm 1\sigma$ and 95% CL intervals.

The contribution from VBF and VH signal production to the inclusive selection gives a small bias when relating the observed event yield ratio

$R_{W/Z}$ to the ratio of the couplings. To assess this effect, a more complex parametrization of the expected signal yield is also used, including the dependency of the VBF and VH cross sections on the W and Z couplings to the Higgs boson. The results extracted using this parameterization agree with the ones from the simpler analysis within 2%, indicating that the bias is negligible compared to the current precision on the measurement.

The measurement, albeit with limited statistical precision, are therefore consistent with the predictions from custodial symmetry.

9.4.2. Probing couplings to fermions and bosons

The ratio of the couplings of the Higgs boson to the electroweak gauge bosons is expected to be equal to the SM prediction also in most beyond-SM scenarios. Conversely, the couplings of the Higgs boson to the fermion sector, and the absolute value of the boson couplings, can exhibit larger deviations in those scenarios.

In order to test for these deviations, a fit to the data is performed using a model that assumes custodial symmetry ($R_{W/Z} = 1$) but allows different couplings for fermions and vector bosons. With the present dataset there is not enough sensitivity to probe separately the couplings to the different fermion types, so the expected signal yields are parametrized in terms of just two modifiers, c_V and c_F, scaling the couplings to the vector bosons and fermions respectively.

A test of the standard model predictions in this context can then be performed by comparing the c_V, c_F values fitted from the data with the $(c_V, c_F) = (1, 1)$ corresponding to a SM Higgs boson. It should however be noted that $(1, 1)$ is the only point at which the predictions from this model are fully consistent, since SM relations are used throughout the analyses when modelling the signal, and also in the computation of higher order corrections to the production cross section and decay widths. Therefore the fit for c_V, c_F is not yet a measurement of the couplings, but can nonetheless yield indications about possible deviations in the coupling structure of the observed Higgs-like candidate.

In the narrow width approximation, which is valid for a light SM-like Higgs boson, the production and decay processes factorize, and the expected signal yields scale as function the partial widths for the different channels according to

$$N(xx \to H \to yy) \sim \frac{\Gamma_{xx}\,\Gamma_{yy}}{\Gamma_{tot}}. \tag{9.2}$$

where the total width is the sum of the partial widths.

At leading order, all tree-level partial widths scale either as c_V^2 or c_F^2. The partial width Γ_{gg} also scales as c_F^2, the effective coupling of the H to gluons being induced only by loop diagrams involving quarks. The partial width $\Gamma_{\gamma\gamma}$ is also induced via loop diagrams, with the W boson and top quark being the dominant contributors; hence, it scales as $|\alpha\, c_V + \beta\, c_F|^2$, where the factors $\alpha(m_H)$ and $\beta(m_H)$ are taken from predictions for the SM Higgs boson [130]. It is also assumed that there is no other invisible or non-SM decay mode for the Higgs boson.

The model has a two-fold degeneracy since the signs of c_V and c_F can be simultaneously swapped while preserving all the scaling factors; c_V can therefore taken to be positive. The $\Gamma_{\gamma\gamma}$ partial width is also the only place where the relative sign of c_F and c_V affects the results: a positive c_F corresponds to the case where the W and top quark loops interfere destructively, as in the SM, while a negative c_F would formally correspond to a constructive interference, yielding an increased diphoton rate. As the $c_F < 0$ region would correspond to a qualitatively different model than the SM Higgs, and also in unphysical negative values of the fermion masses, in this analysis the constraint $c_F \geq 0$ is also enforced.

The scaling factors for the signal yield in each analyzed production and decay mode as function of c_V and c_F are summarized in Table 9.3. For a light SM-like Higgs boson, the total width is dominated by decay modes scaling as c_F^2, and the W contribution dominates the diphoton partial width, so the signal normalization in the different modes scales approximately as c_V^4/c_F^2 for modes with bosons in the production and in the

Production	Decay	Scaling factor		
gg → H	H → γγ	$c_F^2 \cdot |\alpha\, c_V + \beta\, c_F|^2/\Gamma_{\text{tot}}$	~	c_V^2
VBF	H → γγ	$c_V^2 \cdot |\alpha\, c_V + \beta\, c_F|^2/\Gamma_{\text{tot}}$	~	c_V^4/c_F^2
gg → H	H → WW	$c_F^2 \cdot c_V^2/\Gamma_{\text{tot}}$	~	c_V^2
VBF, VH	H → WW	$c_V^2 \cdot c_V^2/\Gamma_{\text{tot}}$	~	c_V^4/c_F^2
gg → H	H → ZZ	$c_F^2 \cdot c_V^2/\Gamma_{\text{tot}}$	~	c_V^2
VBF, VH	H → ZZ	$c_V^2 \cdot c_V^2/\Gamma_{\text{tot}}$	~	c_V^4/c_F^2
ttH	H → b$\bar{\text{b}}$	$c_F^2 \cdot c_F^2/\Gamma_{\text{tot}}$	~	c_F^2
VH	H → b$\bar{\text{b}}$	$c_V^2 \cdot c_F^2/\Gamma_{\text{tot}}$	~	c_V^2
gg → H	H → ττ	$c_F^2 \cdot c_F^2/\Gamma_{\text{tot}}$	~	c_F^2
VBF, VH	H → ττ	$c_V^2 \cdot c_F^2/\Gamma_{\text{tot}}$	~	c_V^2

Table 9.3. Scaling factors for the signal yields in the different production and decay modes considered. The approximate dependency of the scale factor is also given taking into account that Γ_{tot} at low mass is dominated by $\Gamma_{\text{b}\bar{\text{b}}} \sim c_F^2$ and that $\alpha \gg \beta$ in the diphoton decay. The ZZ analysis is done inclusively, and is therefore sensitive only to the total production cross section.

decay, c_F^2 for modes with fermions in the production and decay, and c_V^2 for the other modes. As eventually the most sensitive modes are the ones scaling as c_V^2, the c_V parameter is expected to be probed with a higher accuracy than c_F.

The two-dimensional likelihood scan and the 68% and 95% confidence regions for c_V and c_F are shown in Figure 9.20: the fit to the data yields $(c_V, c_F) \sim (1.0, 0.5)$, where the c_F value smaller than unity is expected from the excess in the VBF H $\rightarrow \gamma\gamma$ mode and the deficit in the fermion modes, but overall the fit is compatible with the SM at 95%. From this analysis it can also be noted that a purely fermiophobic Higgs boson, $c_F = 0$, is disfavoured at more than 99% from the present data.

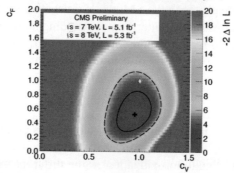

Figure 9.20. Likelihood scan as function of c_V, c_F for the combination of all the CMS searches. The diamond at $(1, 1)$ corresponds to the SM Higgs predictions. The best fit point is indicated by the black cross, and the 68% and 95% CL contours by the solid and dashed lines respectively.

In order to better visualize the contribution of the individual channels to the result, the same c_V, c_F analysis is also performed separately for each Higgs boson decay mode (Figure 9.21). As expected, the fermionic analyses modes constrain c_F at small values of c_V (through tt̄H and gg $\rightarrow$ H $\rightarrow \tau\tau$), and disfavour large values of c_V and c_F simultaneously, especially for the H $\rightarrow \tau\tau$ case, from the VBF and VH production topologies. The constraint on c_V weakens however for $c_F/c_V \lesssim 0.5$, below which the decays mediated by vector bosons become the main contribution to the total width, so that the signal yield becomes approximately proportional to c_F^2 instead of c_V^2 even in those production modes.

The H $\rightarrow$ WW and H $\rightarrow$ ZZ modes both set a constraint on c_V, as gluon-fusion production dominates the rate; in addition, H $\rightarrow$ WW also excludes small values of c_F for moderate c_V, that would imply a larger-than-SM signal in the VBF topology, not observed. Finally, H $\rightarrow \gamma\gamma$ favours either $c_V > 1$ or $c_F < 1$ values, to accommodate for the excess above the SM Higgs boson predictions, especially in the VBF mode. Both H $\rightarrow$ WW and H $\rightarrow \gamma\gamma$, however, also exclude the hypothesis of vanishing fermion couplings, from the interplay of the signal yields measured in VBF and untagged topologies.

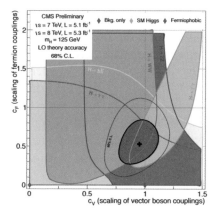

Figure 9.21. 68%CL contours in the (c_V, c_F) plane for the combination of the CMS Higgs searches separately in the five decay modes analyzed and for their combination. For the combination, the 95% CL contour is also shown. The predictions for a SM Higgs boson $(1, 1)$, a purely fermiophobic Higgs boson $(1, 0)$ and for the background only hypothesis $(0, 0)$ are also shown as diamonds.

One-dimensional likelihood scans can also be made fixing c_F or c_V to unity, to determine the allowed range for the other parameter; this is especially well motivated in the c_V case, as several beyond-SM scenarios do not modify the couplings of the Higgs-like boson to the W and Z bosons (Figure 9.22). The 95% CL intervals extracted from this scan are are $0.7 < c_V < 1.2$ and $0.3 < c_F < 1.0$, also compatible with the SM prediction at 95% level.

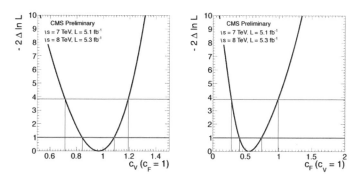

Figure 9.22. Likelihood scan as function of c_V, c_F for the combination of all the CMS searches. The diamond at $(1, 1)$ corresponds to the SM Higgs predictions. The best fit point is indicated by the black cross, and the 68% and 95% CL contours by the solid and dashed lines respectively.

Concluding, the couplings of the Higgs-like candidate observed by CMS have been probed to the extent possible from the available data, and were found to be fairly compatible with the SM Higgs predictions. However, given the limited sensitivity, deviations of up to 50% cannot yet be excluded, and thus there is still room for new physics scenarios which do not depart too much from the standard model (*e.g.* supersymmetry close to the decoupling limit).

Conclusions

The search for a Higgs boson, and more generally the investigation of the electroweak symmetry breaking, has been one of the major goals of the LHC physics program since its design phase.

Thanks to the excellent performance of the accelerator and of the CMS and ATLAS experiments, the data collected in the first three years of running has finally provided a first answer to this long-standing questions. A new heavy boson has been discovered, compatible with the predictions of a SM Higgs boson within the current, fairly large, uncertainties.

The 2010 dataset, while too limited in terms of integrated luminosity for the SM Higgs boson searches, has allowed both experiments to fully commission their physics capabilities by performing measurements of several key standard model processes such as the production of W and Z bosons in association with hadronic jets, diboson and top quark pairs. The 2011 run, corresponding to an increase in integrated luminosity by two orders of magnitude, has allowed the two experiments to reach the sensitivity to the SM Higgs boson, and to restrict its possible existence to a narrow region in the mass range, also favoured by electroweak precision observables, where first hints of a signal have been detected. Eventually these hints have been turned into a full observation with the further doubling of the integrated luminosity with the summer 2012 dataset.

The observation of this new Higgs-like state at a mass of about $125 \, \text{GeV}/c^2$ has started a new phase in the Higgs boson physics program: while the search for possible additional states continues, as many beyond-SM scenarios predict multiple Higgs-like bosons, the foremost priority is now the measurement of the properties of this new state to ascertain its compatibility with the standard model predictions. In that respect, the CMS experiment has already started the process by releasing first results on the mass of this particle, the compatibility between observed and predicted signal yields in the different final state topologies, and tests for deviations in the couplings to electroweak gauge bosons and fermions. Preparations for the measurements of the spin and parity of the boson have also been started, but the current data sample is too limited to provide valuable information. So far, all results are in agreement with the standard model predictions within the uncertainties, but further data from the full 2012 run and beyond will be needed to improve the accuracy of the results.

References

[1] S. GLASHOW, *Partial symmetries of weak interactions*, Nucl.Phys. **22** (1961), 579–588. doi:10.1016/0029-5582(61)90469-2

[2] S. WEINBERG, *A model of leptons*, Phys.Rev.Lett. **19** (1967), 1264–1266. doi:10.1103/PhysRevLett.19.1264

[3] A. SALAM (ed.), "Elementary Particle Physics", Almquvist and Wiksell, Stockholm, 1968, p. 367.

[4] F. ENGLERT and R. BROUT, *Broken symmetry and the mass of gauge vector mesons*, Phys. Rev. Lett. **13** (1964), 321–323. doi:10.1103/PhysRevLett.13.321

[5] P. W. HIGGS, *Broken symmetries, massless particles and gauge fields*, Phys. Lett. textbf12 (1964), 132–133. doi:10.1016/0031-9163(64)91136-9

[6] P. W. HIGGS, *Broken symmetries and the masses of gauge bosons*, Phys. Rev. Lett. **13** (1964), 508–509. doi:10.1103/PhysRevLett.13.508

[7] G. GURALNIK, C. HAGEN and T. W. B. KIBBLE, *Global conservation laws and massless particles*, Phys. Rev. Lett. **13** (1964), 585–587. doi:10.1103/Phys RevLett.13.585

[8] P. W. HIGGS, *Spontaneous symmetry breakdown without massless bosons*, Phys. Rev. **145** (1966), 1156–1163. doi:10.1103/PhysRev.145.1156

[9] T. W. B. KIBBLE, *Symmetry breaking in non-Abelian gauge theories*, Phys. Rev. **155** (1967), 1554–1561. doi:10.1103/PhysRev.155.1554

[10] S. DITTMAIER, C. MARIOTTI *et al.*, LHC Higgs Cross Section Working Group, In: "Handbook of LHC Higgs Cross Sections: 1. Inclusive Observables", CERN-2011-002 (2011). arXiv:1101.0593

[11] S. DITTMAIER, C. MARIOTTI *et al.*, LHC Higgs Cross Section Working Group, In: "Handbook of LHC Higgs Cross Sections: 2. Differential Distributions", CERN-2011-002 (2012). arXiv:1201.3084

[12] ALEPH, DELPHI, L3, OPAL COLLABORATIONS and LEP
     WORKING GROUP FOR HIGGS BOSON SEARCHES, *Search for the
     standard model higgs boson at LEP*, Phys. Lett. B **565** (2003), 61–
     75. doi:10.1016/S0370-2693(03)00614-2

[13] CDF and D0 COLLABORATIONS, *Combination of Tevatron
     Searches for the Standard Model Higgs Boson in the WW Decay
     Mode*, Phys. Rev. Lett. **104** (2010), 061802.
     doi:10.1103/PhysRevLett.104.061802

[14] TEVNPH (TEVATRON NEW PHENOMINA AND HIGGS WORK-
     ING GROUP), CDF and D0 COLLABORATION, *Combined CDF
     and D0 search for standard model Higgs boson production with up
     to 10.0 fb^{-1} of data*. arXiv:1203.3774

[15] ALEPH, CDF, D0, DELPHI, L3, OPAL, SLD COLLABORA-
     TIONS, and LEP ELECTROWEAK WORKING GROUP, the Tevatron
     Electroweak Working Group, and the SLD Electroweak and Heavy
     Flavour Groups, *Precision Electroweak Measurements and Con-
     straints on the Standard Model*, CERN-PH-EP-2010-095 (Decem-
     ber, 2010). arXiv:1012.2367. A recent, unpublished, limit can be
     found at http://lepewwg.web.cern.ch/LEPEWGG/

[16] L. EVANS and P. BRYANT (eds.), "LHC Machine", JINST **3** (2008),
     S08001. doi:10.1088/1748-0221/3/08/S08001

[17] CMS COLLABORATION, *The CMS experiment at the CERN LHC*,
     JINST **03** (2008) S08004 (2008).
     doi:10.1088/1748-0221/3/08/S08004

[18] ATLAS COLLABORATION, *The ATLAS Experiment at the
     CERN Large Hadron Collider"*, JINST **3** (2008) S08003.
     doi:10.1088/1748-0221/3/08/S08003

[19] CMS COLLABORATION, *Combined results of searches for the
     standard model Higgs boson in* pp *collisions at* $\sqrt{s} = 7$ TeV, Phys.
     Lett. **B710** (2012), 26–48.
     http://dx.doi.org/10.1016/j.physletb.2012.02.064
     doi:10.1016/j.physletb.2012.02.064

[20] ATLAS COLLABORATION, *Combined search for the Standard
     Model Higgs boson using up to 4.9* fb^{-1} *of* pp *collision data at*
     $\sqrt{s} = 7$ TeV *with the ATLAS detector at the LHC*, Phys. Lett.
     B710 (2012), 49–66. arXiv:1202.1408

[21] CMS COLLABORATION, *Combined results of searches for a Higgs
     boson in the context of the standard model and beyond-standard
     models*, CMS Physics Analysis Summary (2012). **HIG-12-008**

[22] ATLAS COLLABORATION, *An update to the combined search for
     the Standard Model Higgs boson with the* ATLAS *detector at the*

LHC *using up to* $4.9\,\mathrm{fb}^{-1}$ *of* pp *collision data at* $\sqrt{s} = 7$ TeV, ATLAS Note (2012). **ATLAS-CONF-2012-019**

[23] CMS COLLABORATION, *Observation of a new boson at a mass of* 125 GeV *with the* CMS *experiment at the* LHC, Phys. Lett. **B** (2012). doi:10.1016/j.physletb.2012.08.021

[24] ATLAS COLLABORATION, *Observation of a new particle in the search for the Standard Model Higgs boson with the* ATLAS *detector at the* LHC, Phys. Lett. **B** (2012). doi:10.1016/j.physletb.2012.08.020

[25] W. STIRLING, "MSTW Plots", private communication.

[26] R. FRUHWIRTH, *Application of Kalman filtering to track and vertex fitting*, Nucl. Instrum. Meth. **A262** (1987), 444–450. doi:10.1016/0168-9002(87)90887-4

[27] W. ADAM, B. MANGANO, T. SPEER et al., *Track reconstruction in the CMS tracker*, CMS NOTE (2006). **2006/041**

[28] CMS COLLABORATION, *Measurement of Tracking Efficiency*, CMS Physics Analysis Summary, (2010). **TRK-10-002**

[29] CMS COLLABORATION, *Measurement of the Inclusive* W *and* Z *Production Cross Sections in* pp *Collisions at* $\sqrt{(s)} = 7$ TeV, JHEP **1110** (2011), 132. doi:10.1007/JHEP10(2011)132

[30] CMS COLLABORATION, *Performance of CMS muon reconstruction in* pp *collisions at* $\sqrt{(s)} = 7$ TeV, submitted to JINST arXiv:1206.4071

[31] CMS COLLABORATION, *Tracking and Vertexing Results in First 7 TeV Collisions*, CMS Physics Analysis Summary (2010). **TRK-10-005**

[32] K. ROSE, *Deterministic annealing for clustering, compression, classification, regression, and related optimization problems*, Proceedings of the IEEE **86** (nov, 1998), 2210 –2239. doi:10.1109/5.726788

[33] E. CHABANAT and N. ESTRE, *Deterministic annealing for vertex finding at CMS*, Proceeedings of CHEP 2004 (2005), 287–290.

[34] R. FRUHWIRTH, W. WALTENBERGER and P. VANLAER, *Adaptive vertex fitting*, J. Phys. G **34** (2007), N343. doi:10.1088/0954-3899/34/12/N01

[35] CMS COLLABORATION, b-*jet identification in the CMS experiment*, CMS Physics Analysis Summary (2012). **BTV-11-004**

[36] CMS Collaboration, "Measurement of btagging efficiency using t$\bar{\mathrm{t}}$ events", *CMS Physics Analysis Summary* (2012). **BTV-11-003**

[37] CMS COLLABORATION, *Performance of CMS Muon Reconstruction in Cosmic-Ray Events*, JINST **5** (2010), T03022. doi:10.1088/1748-0221/5/03/T03022

[38] CMS COLLABORATION, *Performance of muon identification in* pp *collisions at* $\sqrt{s} = 7$ TeV, CMS Physics Analysis Summary (2010). **MUO-10-002**

[39] CMS COLLABORATION, *Commissioning of the particle-flow event reconstruction with leptons from J/Psi and W decays at* 7 TeV, CMS Physics Analysis Summary (2010). **PFT-10-003**

[40] M. CACCIARI and G. P. SALAM, *Pileup subtraction using jet areas*, Phys. Lett. **B659** (2008), 119–126. doi:10.1016/j.physletb.2007.09.077

[41] M. CACCIARI, G. P. SALAM and G. SOYEZ, *The Catchment Area of Jets*, JHEP **04** (2008), 005. doi:10.1088/1126-6708/2008/04/005

[42] CMS COLLABORATION, *Studies of tracker material*, CMS Physics Analysis Summary (2010). **TRK-10-003**

[43] W. ADAM, R. FRUHWIRTH, A. STRANDLIE *et al.*, *Reconstruction of electrons with the Gaussian sum filter in the CMS tracker at LHC*, J. Phys, G: Nucl. Part. Phys **31** (2005), N5–N20. doi:10.1088/0954-3899/31/9/N01

[44] S. BAFFIONI, C. CHARLOT, F. FERRI *et al.*, *Electron reconstruction in CMS*, Eur. Phys. J. **C49** (2007), 1099–1116. doi:10.1140/epjc/s10052-006-0175-5

[45] CMS COLLABORATION, *Electron reconstruction and identification at* $\sqrt{s} = 7$ TeV, CMS Physics Analysis Summary (2010). **EGM-10-004**

[46] CMS COLLABORATION, *Photon reconstruction and identification at* $\sqrt{s} = 7$ TeV, CMS Physics Analysis Summary (2010). **EGM-10-005**

[47] CMS COLLABORATION, *Particle-Flow Event Reconstruction in CMS and Performance for Jets, Taus, and MET*, CMS Physics Analysis Summary (2009). **PFT-09-001**

[48] CMS COLLABORATION, *Commissioning of the particle-flow event reconstruction with the first LHC collisions recorded in the CMS detecto*, CMS Physics Analysis Summary (2010). **PFT-10-001**

[49] CMS COLLABORATION, *Commissioning of the Particle-Flow reconstruction in Minimum-Bias and Jet Events from* pp *Collisions at* 7 TeV, CMS Physics Analysis Summary (2010). **PFT-10-002**

[50] M. CACCIARI, G. P. SALAM and G. SOYEZ, *The anti-kt jet clustering algorithm*, JHEP **04** (2008), 063. doi:10.1088/1126-6708/2008/04/063

[51] CMS COLLABORATION, *Determination of jet energy calibration and transverse momentum resolution in CMS*, JINST **06** (2011), P11002. doi:10.1088/1748-0221/6/11/P11002

[52] C. COLLABORATION, *Missing transverse energy performance of the CMS detector*, JINST **6** (2011), P09001.
doi:10.1088/1748-0221/6/09/P09001

[53] C. COLLABORATION, *Performance of tau-lepton reconstruction and identification in CMS*, JINST **7** (2012), P01001.
arXiv:1109.6034

[54] PARTICLE DATA GROUP COLLABORATION, *Review of particle physics*, J. Phys. G **37** (2010), 075021.
doi:10.1088/0954-3899/37/7A/075021

[55] CMS COLLABORATION, *CMS technical design report, volume II: Physics performance*, J. Phys. G **34** (2007), 995–1579.
doi:10.1088/0954-3899/34/6/S01

[56] CMS COLLABORATION, *Measurement of the Inclusive Z Cross Section via Decays to Tau Pairs in* pp *Collisions at* $\sqrt{s} = 7$ TeV, JHEP **1108** (2011), 117. doi:10.1007/JHEP08(2011)117

[57] S. FRIXIONE, P. NASON and C. OLEARI, *Matching NLO QCD computations with parton shower simulations: the POWHEG method*, JHEP **11** (2007), 070. doi:10.1088/1126-6708/2007/11/070

[58] S. ALIOLI, P. NASON, C. OLEARI *et al.*, *NLO Higgs boson production via gluon fusion matched with shower in POWHEG*, JHEP **04** (2009), 002. doi:10.1088/1126-6708/2009/04/002

[59] P. NASON and C. OLEARI, *NLO Higgs boson production via vector-boson fusion matched with shower in POWHEG*, JHEP **02** (2010), 037. doi:10.1007/JHEP02(2010)037

[60] S. ALIOLI, P. NASON, C. OLEARI *et al.*, *NLO vector-boson production matched with shower in POWHEG*, JHEP **0807** (2008), 060.
doi:10.1088/1126-6708/2008/07/060

[61] S. ALIOLI, P. NASON, C. OLEARI *et al.*, *NLO single-top production matched with shower in POWHEG: s- and t-channel contributions*, JHEP **0909** (2009), 111.
doi:10.1088/1126-6708/2009/09/111, 10.1007/JHEP02(2010)011

[62] E. RE, *Single-top Wt-channel production matched with parton showers using the POWHEG method*, Eur. Phys. J. **C71** (2011), 1547. doi:10.1140/epjc/s10052-011-1547-z

[63] J. ALWALL, P. DEMIN, S. DE VISSCHER *et al.*, *MadGraph/Mad Event v4: the new web generation*, JHEP **09** (2007), 028.
doi:10.1088/1126-6708/2007/09/028

[64] T. BINOTH, M. CICCOLINI, N. KAUER *et al.*, *Gluon-induced W-boson pair production at the LHC*, JHEP **12** (2006), 046.
doi:10.1088/1126-6708/2006/12/046

[65] T. SJOSTRAND, S. MRENNA and P. Z. SKANDS, *PYTHIA 6.4 Physics and Manual*, JHEP **0605** (2006), 026. doi:10.1088/1126-6708/2006/05/026

[66] S. FRIXIONE and B. R. WEBBER, *Matching NLO QCD computations and parton shower simulations*, JHEP **0206** (2002), 029. arXiv:hep-ph/0204244

[67] G. CORCELLA, I. KNOWLES, G. MARCHESINI *et al.*, *HERWIG 6: An Event generator for hadron emission reactions with interfering gluons (including supersymmetric processes)*, JHEP **0101** (2001), 010. ARXIV:HEP-PH/0011363

[68] S. JADACH, J. H. KUHN and Z. WĄS, *TAUOLA - a library of Monte Carlo programs to simulate decays of polarized tau leptons*, COMPUTER PHYSICS COMMUNICATIONS **64** (1991), 275. doi:10.1016/0010-4655(91)90038-M

[69] GEANT4 COLLABORATION, *GEANT4: A Simulation toolkit*, Nucl. Instrum. Meth. **A506** (2003), 250. doi:10.1016/S0168-9002(03)01368-8

[70] R. GAVIN, Y. LI, F. PETRIELLO *et al.*, *FEWZ 2.0: A code for hadronic Z production at next-to-next-to-leading order*, Computer Physics Communications **182** (2011), 2388–2403. doi:10.1016/j.cpc.2011.06.008

[71] M. DITTMAR and H. K. DREINER, *How to find a Higgs boson with a mass between 155 GeV and 180 GeV at the CERN LHC*, Phys. Rev. **D55** (1996), 167. doi:10.1103/PhysRevD.55.167

[72] A. HOCKER, J. STELZER, F. TEGENFELDT *et al.*, *TMVA - toolkit for multivariate data analysis*, PoS **ACAT** (2007), 040. arXiv:physics/0703039

[73] J. M. CAMPBELL, R. ELLIS and C. WILLIAMS, *Gluon-Gluon contributions to W^+W^- production and higgs interference effects*, JHEP **1110** (2011), 005. doi:10.1007/JHEP10(2011)005

[74] CDF COLLABORATION, *Search for a higgs boson decaying to two W bosons at CDF*, Phys. Rev. Lett. **102** (2009), 021802. doi:10.1103/PhysRevLett.102.021802

[75] CMS COLLABORATION, *Absolute calibration of the CMS luminosity measurement: summer 2011 update*, CMS Physics Analysis Summary (2011). **EWK-11-001**

[76] ATLAS COLLABORATION, CMS COLLABORATION and LHC HIGGS COMBINATION GROUP, *Procedure for the LHC higgs boson search combination in summer 2011* (August, 2011). *ATL-PHYS-PUB-2011-011, CMS NOTE-2011/005*

[77] J. M. CAMPBELL, R. ELLIS and C. WILLIAMS, *Hadronic production of a higgs boson and two jets at next-to-leading order*, Phys. Rev. **D81** (2010), 074023. doi:10.1103/PhysRevD.81.074023

[78] M. BOTJE *et al.*, *The PDF4LHC working group interim recommendations*. arXiv:1101.0538

[79] S. ALEKHIN *et al.*, *The PDF4LHC working group interim report*. arXiv:1101.0536

[80] H.-L. LAI, M. GUZZI, J. HUSTON *et al.*, *New parton distributions for collider physics*, Phys. Rev. **D82** (2010), 074024. doi:10.1103/PhysRevD.82.074024

[81] A. D. MARTIN *et al.*, *Parton distributions for the LHC*, Eur. Phys. J. C **63** (2009), 189–285. doi:10.1140/epjc/s10052-009-1072-5

[82] NNPDF COLLABORATION, *Impact of heavy quark masses on Parton distributions and LHC phenomenology*. arXiv:1101.1300

[83] LHC HIGGS CROSS SECTION WORKING GROUP, "Heavy Higgs Line Shape".

[84] J. M. BUTTERWORTH, J. R. FORSHAW and M. H. SEYMOUR, *Multiparton interactions in photoproduction at HERA*, Z. Phys. **C72** (1996), 637–646. doi:10.1007/s002880050286

[85] J. M. CAMPBELL and F. TRAMONTANO, *Next-to-leading order corrections to Wt production and decay*, Nucl. Phys. **B726** (2005), 109–130. doi:10.1016/j.nuclphysb.2005.08.015

[86] C. D. WHITE, S. FRIXIONE, E. LAENEN *et al.*, *Isolating Wt production at the LHC*, JHEP **0911** (2009), 074. doi:10.1088/1126-6708/2009/11/074

[87] Y. GAO, A. V. GRITSAN, Z. GUO *et al.*, *Spin determination of single-produced resonances at hadron colliders*, Phys. Rev. **D81** (2010), 075022. doi:10.1103/PhysRevD.81.075022

[88] A. DE RUJULA, J. LYKKEN, M. PIERINI *et al.*, *Higgs look-alikes at the LHC*, Phys. Rev. D **82** (2010), 013003. doi:10.1103/PhysRevD.82.013003

[89] S. CHOI, D. MILLER, M. MUHLLEITNER *et al.*, *Identifying the Higgs spin and parity in decays to Z pairs*, Phys. Lett. B **553** (2003), 61. doi:10.1016/S0370-2693(02)03191-X

[90] J. M. CAMPBELL and R. K. ELLIS, *MCFM for the Tevatron and the LHC*, Nucl. Phys. Proc. Suppl. **205** (2010), 10. doi:10.1016/j.nuclphysbps.2010.08.011

[91] J. M. CAMPBELL, R. ELLIS and C. WILLIAMS, *Vector boson pair production at the LHC*, JHEP **07** (2011), 018. doi:10.1007/JHEP07(2011)018

[92] T. MELIA, P. NASON, R. RONTSCH et al., W + W−, WZ and ZZ production in the POWHEG BOX, JHEP 11 (2011), 078. doi:10.1007/JHEP11(2011)078

[93] T. BINOTH, N. KAUER and P. MERTSCH, Gluon-induced QCD corrections to pp → ZZ → $\ell\bar{\ell}\ell'\bar{\ell}'$, Proceedings of the XVI Int. Workshop on Deep-Inelastic Scattering and Related Topics (DIS'07) (2008). doi:10.3360/dis.2008.142

[94] C. COLLABORATION, Search for the standard model Higgs boson decaying into two photons in pp collisions at $\sqrt{(s)} = 7$ TeV, Phys. Lett. B710 (2012), 403–425. doi:10.1016/j.physletb.2012.03.003

[95] D. L. RAINWATER, R. SZALAPSKI and D. ZEPPENFELD, Probing color singlet exchange in Z+ two jet events at the CERN LHC, Phys. Rev. D54 (1996), 6680–6689. doi:10.1103/PhysRevD.54.6680

[96] CMS COLLABORATION, Search for neutral MSSM higgs bosons decaying to tau pairs in pp collisions at $\sqrt{s} = 7$ TeV, Phys. Rev. Lett. 106 (2011), 231801. doi:10.1103/PhysRevLett.106.231801

[97] D. L. RAINWATER, D. ZEPPENFELD and K. HAGIWARA, Searching for H → ττ in weak boson fusion at the CERN LHC, Phys. Rev. D59 (1998), 014037. doi:10.1103/PhysRevD.59.014037

[98] CMS COLLABORATION, Combination of top pair production cross section measurements, CMS Physics Analysis Summary (2011). TOP-11-024

[99] J. M. BUTTERWORTH, A. R. DAVISON, M. RUBIN et al., Jet substructure as a new Higgs search channel at the LHC, Phys.Rev.Lett. 100 (2008), 242001. doi:10.1103/PhysRevLett.100.242001

[100] J. GALLICCHIO and M. D. SCHWARTZ, Seeing in Color: Jet Superstructure, Phys. Rev. Lett. 105 (2010), 022001. doi:10.1103/PhysRevLett.105.022001

[101] CTEQ COLLABORATION, Coordinated Theoretical-Experimental project on QCD.

[102] R. D. COUSINS and V. L. HIGHLAND, Incorporating systematic uncertainties into an upper limit, Nucl. Instrum. Meth. A320 (1992), 331–335. Revised version. doi:10.1016/0168-9002(92)90794-5

[103] G. COWAN et al., Asymptotic formulae for likelihood-based tests of new physics, Eur. Phys. J. C 71 (2011), 1–19. doi:10.1140/epjc/s10052-011-1554-0

[104] E. GROSS and O. VITELLS, Trial factors for the look elsewhere effect in high energy physics, Eur. Phys. J. C 70 (2010), 525–530. doi:10.1140/epjc/s10052-010-1470-8

[105] J. NEYMAN, Outline of a theory of statistical estimation based on the classical theory of probability, Philosophical Transactions of

the Royal Society of London. Series A, Mathematical and Physical Sciences **236** (1937), no. 767, pp. 333–380.

[106] R. D. C. GARY J. FELDMAN, *Unified approach to the classical statistical analysis of small signals*, Phys. Rev. **D57** (1998), no. 7, 3873–3889. doi:10.1103/PhysRevD.57.3873

[107] A. L. READ, "Modified frequentist analysis of search results (the CLs method)", CERN Yellow Report (2000), 81. **CERN-2000-005**

[108] T. JUNK, *Confidence level computation for combining searches with small statistics*, Nucl. Instrum. Meth. A **434** (1999), 435–443. doi:10.1016/S0168-9002(99)00498-2

[109] A. L. READ, *Presentation of search results: The CL(s) technique*, J. Phys. G: Nucl. Part. Phys. **G28** (2002), 2693–2704. doi:10.1088/0954-3899/28/10/313

[110] S. S. WILKS, *The large-sample distribution of the likelihood ratio for testing composite hypotheses*, Ann. Math. Statist. **9** (1938), 60–62. doi:10.1214/aoms/1177732360

[111] CMS COLLABORATION, *Observation of a new boson with a mass near 125 GeV*, CMS Physics Analysis Summary (2012). **HIG-12-020**

[112] CMS COLLABORATION, *Search for the standard model Higgs boson produced in association with W or Z bosons, and decaying to bottom quarks*, CMS Physics Analysis Summary (2012). **HIG-12-019**

[113] CMS COLLABORATION, *Search for Higgs production in association with top quark pairs in* pp *collisions*, CMS Physics Analysis Summary (2012). **HIG-12-025**

[114] CMS COLLABORATION, *Search for neutral Higgs bosons decaying to tau pairs in* pp *collisions*, CMS Physics Analysis Summary (2012). **HIG-12-018**

[115] CMS COLLABORATION, *Search for the standard model Higgs boson in associated WH production in the* $e\mu\tau$ *and* $\mu\mu\tau$ *final states*, CMS Physics Analysis Summary (2012). **HIG-12-006**

[116] CMS COLLABORATION, *Search for the standard model Higgs boson decaying into* $\tau\tau$ *and WW associated with Z*, CMS Physics Analysis Summary (2012). **HIG-12-012**

[117] CMS COLLABORATION, *Evidence for a new state decaying into two photons in the search for the standard model Higgs boson in* pp *collisions*, CMS Physics Analysis Summary (2012). **HIG-12-015**

[118] CMS COLLABORATION, *Search for the Higgs Boson Decaying to* W^+W^- *in the Fully Leptonic Final State*, CMS Physics Analysis Summary (2011). **HIG-11-024**

[119] CMS COLLABORATION, *Search for the standard model Higgs boson decaying to* W^+W^- *in the fully leptonic final state in* pp *collisions at* $\sqrt{s} = 8$ TeV, CMS Physics Analysis Summary (2012). **HIG-12-017**

[120] CMS COLLABORATION, *Search for the Standard Model Higgs boson in the* H $\rightarrow$ WW $\rightarrow$ lνjj *decay channel*, CMS Physics Analysis Summary (2012). **HIG-12-003**

[121] CMS COLLABORATION, *Search for the Standard Model Higgs boson in the* $H \rightarrow WW \rightarrow l\nu jj$ *decay channel in* pp *collisions at* $\sqrt{s} = 8$ TeV, CMS Physics Analysis Summary (2012). **HIG-12-021**

[122] CMS COLLABORATION, *Study of associated Higgs boson* (WH) *Production in the three leptons final state at* 7 TeV, CMS Physics Analysis Summary (2011). **HIG-11-034**

[123] CMS COLLABORATION, *Search for associated Higgs production* (VH) *with* H $\rightarrow$ W^+W^- $\rightarrow$ lνlν *and hadronic* V *decay in* pp *collisions at* $\sqrt{s} = 7$ TeV, CMS Physics Analysis Summary (2012). **HIG-12-014**

[124] CMS COLLABORATION, *Evidence for a new state in the search for the standard model Higgs boson in the* H $\rightarrow$ ZZ $\rightarrow$ 4l *channel in* pp *collisions at* $\sqrt{s} = 7$ *and* 8 *TeV*, CMS Physics Analysis Summary (2012). **HIG-12-016**

[125] CMS COLLABORATION, *Search for a Higgs boson in the decay channel* H $\rightarrow$ ZZ $\rightarrow$ 2lqq, CMS Physics Analysis Summary (2011). **HIG-11-027**

[126] CMS COLLABORATION, *Search for the standard model Higgs boson in the* $H \rightarrow ZZ \rightarrow 2l2\nu$ *channel in* pp *collisions at* $\sqrt{s} = 7$ *and* 8 TeV, CMS Physics Analysis Summary (2012). **HIG-12-023**

[127] L. D. LANDAU, Dokl. Akad. Nauk **60** (1948), 207.

[128] C. N. YANG, *Selection Rules for the Dematerialization of a Particle into Two Photons*, Phys. Rev. **77** (Jan, 1950), 242–245. doi:10.1103/PhysRev.77.242

[129] E. BAGNASCHI, G. DEGRASSI, P. SLAVICH *et al.*, *Higgs production via gluon fusion in the POWHEG approach in the SM and in the MSSM*, JHEP **1202** (2012), 088. doi:10.1007/JHEP02(2012)088

[130] SPIRA, M. *et al.*, *Higgs boson production at the LHC*, Nucl. Phys. B **453** (1995), 17–82. doi:10.1016/0550-3213(95)00379-7

THESES

This series gathers a selection of outstanding Ph.D. theses defended at the Scuola Normale Superiore since 1992.

Published volumes

1. F. COSTANTINO, *Shadows and Branched Shadows of 3 and 4-Manifolds*, 2005. ISBN 88-7642-154-8

2. S. FRANCAVIGLIA, *Hyperbolicity Equations for Cusped 3-Manifolds and Volume-Rigidity of Representations*, 2005. ISBN 88-7642-167-x

3. E. SINIBALDI, *Implicit Preconditioned Numerical Schemes for the Simulation of Three-Dimensional Barotropic Flows*, 2007. ISBN 978-88-7642-310-9

4. F. SANTAMBROGIO, *Variational Problems in Transport Theory with Mass Concentration*, 2007. ISBN 978-88-7642-312-3

5. M. R. BAKHTIARI, *Quantum Gases in Quasi-One-Dimensional Arrays*, 2007. ISBN 978-88-7642-319-2

6. T. SERVI, *On the First-Order Theory of Real Exponentiation*, 2008. ISBN 978-88-7642-325-3

7. D. VITTONE, *Submanifolds in Carnot Groups*, 2008. ISBN 978-88-7642-327-7

8. A. FIGALLI, *Optimal Transportation and Action-Minimizing Measures*, 2008. ISBN 978-88-7642-330-7

9. A. SARACCO, *Extension Problems in Complex and CR-Geometry*, 2008. ISBN 978-88-7642-338-3

10. L. MANCA, *Kolmogorov Operators in Spaces of Continuous Functions and Equations for Measures*, 2008. ISBN 978-88-7642-336-9

11. M. LELLI, *Solution Structure and Solution Dynamics in Chiral Ytterbium(III) Complexes*, 2009. ISBN 978-88-7642-349-9

12. G. CRIPPA, *The Flow Associated to Weakly Differentiable Vector Fields*, 2009. ISBN 978-88-7642-340-6

13. F. CALLEGARO, *Cohomology of Finite and Affine Type Artin Groups over Abelian Representations*, 2009. ISBN 978-88-7642-345-1

14. G. DELLA SALA, *Geometric Properties of Non-compact C R Manifolds*, 2009. ISBN 978-88-7642-348-2

15. P. BOITO, *Structured Matrix Based Methods for Approximate Polynomial GCD*, 2011. ISBN: 978-88-7642-380-2; e-ISBN: 978-88-7642-381-9

16. F. POLONI, *Algorithms for Quadratic Matrix and Vector Equations*, 2011. ISBN: 978-88-7642-383-3; e-ISBN: 978-88-7642-384-0

17. G. DE PHILIPPIS, *Regularity of Optimal Transport Maps and Applications*, 2013. ISBN: 978-88-7642-456-4; e-ISBN: 978-88-7642-458-8

18. G. PETRUCCIANI, *The Search for the Higgs Boson at CMS*, 2013. ISBN: 978-88-7642-481-6; e-ISBN: 978-88-7642-482-3

Volumes published earlier

H. Y. FUJITA, *Equations de Navier-Stokes stochastiques non homogènes et applications*, 1992.

G. GAMBERINI, *The minimal supersymmetric standard model and its phenomenological implications*, 1993. ISBN 978-88-7642-274-4

C. DE FABRITIIS, *Actions of Holomorphic Maps on Spaces of Holomorphic Functions*, 1994. ISBN 978-88-7642-275-1

C. PETRONIO, *Standard Spines and 3-Manifolds*, 1995. ISBN 978-88-7642-256-0

I. DAMIANI, *Untwisted Affine Quantum Algebras: the Highest Coefficient of* det H_η *and the Center at Odd Roots of 1*, 1996. ISBN 978-88-7642-285-0

M. MANETTI, *Degenerations of Algebraic Surfaces and Applications to Moduli Problems*, 1996. ISBN 978-88-7642-277-5

F. CEI, *Search for Neutrinos from Stellar Gravitational Collapse with the MACRO Experiment at Gran Sasso*, 1996. ISBN 978-88-7642-284-3

A. SHLAPUNOV, *Green's Integrals and Their Applications to Elliptic Systems*, 1996. ISBN 978-88-7642-270-6

R. TAURASO, *Periodic Points for Expanding Maps and for Their Extensions*, 1996. ISBN 978-88-7642-271-3

Y. BOZZI, *A study on the activity-dependent expression of neurotrophic factors in the rat visual system*, 1997. ISBN 978-88-7642-272-0

M. L. CHIOFALO, *Screening effects in bipolaron theory and high-temperature superconductivity*, 1997. ISBN 978-88-7642-279-9

D. M. CARLUCCI, *On Spin Glass Theory Beyond Mean Field*, 1998. ISBN 978-88-7642-276-8

G. Lenzi, *The MU-calculus and the Hierarchy Problem*, 1998.
ISBN 978-88-7642-283-6

R. Scognamillo, *Principal G-bundles and abelian varieties: the Hitchin system*, 1998. ISBN 978-88-7642-281-2

G. Ascoli, *Biochemical and spectroscopic characterization of CP20, a protein involved in synaptic plasticity mechanism*, 1998.
ISBN 978-88-7642-273-7

F. Pistolesi, *Evolution from BCS Superconductivity to Bose-Einstein Condensation and Infrared Behavior of the Bosonic Limit*, 1998.
ISBN 978-88-7642-282-9

L. Pilo, *Chern-Simons Field Theory and Invariants of 3-Manifolds*, 1999.
ISBN 978-88-7642-278-2

P. Aschieri, *On the Geometry of Inhomogeneous Quantum Groups*, 1999. ISBN 978-88-7642-261-4

S. Conti, *Ground state properties and excitation spectrum of correlated electron systems*, 1999. ISBN 978-88-7642-269-0

G. Gaiffi, *De Concini-Procesi models of arrangements and symmetric group actions*, 1999. ISBN 978-88-7642-289-8

N. Donato, *Search for neutrino oscillations in a long baseline experiment at the Chooz nuclear reactors*, 1999. ISBN 978-88-7642-288-1

R. Chirivì, *LS algebras and Schubert varieties*, 2003.
ISBN 978-88-7642-287-4

V. Magnani, *Elements of Geometric Measure Theory on Sub-Riemannian Groups*, 2003. ISBN 88-7642-152-1

F. M. Rossi, *A Study on Nerve Growth Factor (NGF) Receptor Expression in the Rat Visual Cortex: Possible Sites and Mechanisms of NGF Action in Cortical Plasticity*, 2004. ISBN 978-88-7642-280-5

G. Pintacuda, *NMR and NIR-CD of Lanthanide Complexes*, 2004.
ISBN 88-7642-143-2

Fotocomposizione "CompoMat" Loc. Braccone, 02040 Configni (RI) Italy
Finito di stampare nel mese di ottobre 2013
dalla CSR srl, Via di Pietralata, 157, 00158 Roma